Herbert Edward Forrest

The Fauna of Shropshire

Being an Account of All the Mammals, Birds, Reptiles & Fishes....

Herbert Edward Forrest

The Fauna of Shropshire
Being an Account of All the Mammals, Birds, Reptiles & Fishes....

ISBN/EAN: 9783337025861

Printed in Europe, USA, Canada, Australia, Japan

Cover: Foto ©berggeist007 / pixelio.de

More available books at **www.hansebooks.com**

THE

FAUNA OF SHROPSHIRE

BEING AN ACCOUNT OF ALL THE

MAMMALS, BIRDS, REPTILES & FISHES

FOUND IN THE COUNTY OF SALOP.

With an introduction dealing with the physical features of the
County, a copious index, a Chapter on the principal
Naturalists who have done work in connection
with the subject, and a short account of
the Wild Birds Protection Acts.

———

ILLUSTRATED BY PHOTOGRAPHIC PLATES AND PORTRAITS.

———

BY

H. EDWARD FORREST,

Hon. Loc. Sec. Caradoc and Severn Valley Field Club.
Hon. Corresponding Member and late Hon. Sec. Birmingham Nat. Hist.
and Philosophical Society.

———

Shrewsbury :
L. WILDING, CASTLE STREET.

London :
TERRY & COMPY., 6 HATTON GARDEN, E.C.
—
1899.

DEDICATION.

Happy indeed is the naturalist: to him the seasons come round like old friends; to him the birds sing: as he walks along, the flowers stretch out from the hedges, or look up from the ground; and as each year fades away, he looks back on a fresh store of happy memories.

SIR JOHN LUBBOCK.   *The Beauties of Nature*

PREFACE.

THE very hearty support accorded to this book before
its issue seems to imply that the people of Shropshire
take a real interest in local Natural History. I feel, therefore, that any apology for its publication is unnecessary, but
would like to say a few words as to the form in which it
is written. I have explained in the introductions to the
chapters on Mammals and Birds, my reasons for treating the
former at length, and the latter with brevity, and expressed
a hope that the defect in this instance might be remedied in a
subsequent volume. In the present book I have attempted,
firstly, to give local records a prominent position, and to
render the list of species so complete that it may be relied
on as a work of reference in future years. It is only to be
expected that errors should creep into a work of this nature,
but I can assure my readers that no pains have been spared
to eradicate them. Secondly, it has been my aim to give a
complete, accurate, yet readable account of each species, and
as I always feel that the greatest interest attaches to the
common species, I have, as a rule, devoted more space to
these than the rare or extinct ones. If it should be thought
that the space devoted to the Amphibia is unduly large, I
would plead in extenuation that, although so common, these
humble creatures are not studied as they deserve to be, and
I have not met with *any book* that describes with completeness

and accuracy the wondrous series of transformations which they undergo. Finally, may I ask the scientific reader to pardon the use of popular instead of technical terms in the text. Throughout the work I have striven to write in language such as would be understood by the pupils in the upper classes of elementary schools, as I was told by several members of the teaching profession that such a book was greatly needed for use in giving "object lessons." I have, therefore, made arrangements for the issue of an abridged edition, containing all those portions of the text which would be likely to prove useful for teaching purposes. I cannot conclude without expressing my heartfelt gratitude to the many kind friends who have helped in the preparation of this work. The number of those who have contributed notes is so great that I cannot mention them all by name; but my thanks are especially due to Miss Beckwith, who generously placed at my disposal the whole of the correspondence, and books of cuttings, of the late Mr. Beckwith—a perfect mine of information from which I have drawn freely. I would also express my hearty thanks to Rev. J. B. Meredith, Dr. Rope, Dr. Sankey, Mr. J. Steele Elliot, Mr. H. F. Harries, Mr. F. Rawdon Smith, Mr. Brownlow Tower, and others who have revised many of the sheets of the book; to Mr. John Franklin, Mr. R. J. Irwin, Mrs. Rocke, Hon. Frank Hill, and others who have helped in the preparation of the photographic plates; to Mr. Watkin Watkins for his excellent résumé of the Wild Birds Protection Acts; and to Mr. W. E. Edwards who contributed largely to the chapter on Fishes. I lay down the pen with a feeling of relief that my task is done, yet with the pleasant sense that it has brought me into contact, or correspondence, with many naturalists with whom I had not previously been acquainted. In this connection, however,

I hope that the book will lead to further results, for I shall always be glad to hear from anyone who is interested in the subject, and shall take it as a favour if those who detect errors in the text will inform me of the same; or, if those who meet with additional or unrecorded specimens will let me know particulars. I shall also be very pleased, at any time, to aid in identifying rare or doubtful species of birds, etc., or to give any information in my power.

H. E. FORREST.

37 *Castle Street,*
Shrewsbury,
1st *May,* 1899.

CONTENTS.

LIST OF ILLUSTRATIONS.

———

Abbreviations on the Plates.   M.—Male;   F.—Female;   Y.—Young.

———

Frontispiece   .   .   .   . OSPREYS.

To be inserted *instead of* page 8, in Forrest's " Fauna of Shropshire."

ERRATA.

Page 13, line 17 for " chaper " read " chapter."
 45 „ 29 „ " only " „ " chiefly."
 22 „ 25 „ " Serne Abbot " read " Cerne Abbas."
 35 „ 21 „ " lightly " read " slightly."
 62 „ 5, etc. should read " which has several entrances. These all lead through one hole only just large enough," etc.
 66 „ 17 for " June or July " read " March to July."
 71 „ 67 " Murdoch, states that this mouse was plentiful there then."
 „ 30 The mouse does not store up the berries, but bites through one end of the stone, and extracts the kernel, rejecting the outer pulp.
 84 „ 4 for " course " read " coarse."
 86 „ 28 „ " attacks " „ " attack."
 „ 29 „ " Rabbits " „ " Rabbit."
 90 „ 3 „ " white " „ " dark."
 108 „ 11 „ " a few " „ " many."
 118 „ 31 „ " mulitiplied " read " multiplied."
 „ " valgaris " „ " vulgaris."
 120 „ 20, etc. for " Redpole " read " Redpoll."
 125 „ 7 for " *Plectrophanes* " „ " *Plectrophenax.*"
 129 „ 24 „ " but it is said " „ " except."
 130 „ 30 Omit " Bee-eater " paragraph.
 131 „ 5 for " 1889 " read " 1887."
 „ 16 read " Mr. B. Salter says that at Ley Grange."
 134 „ 13 for " *lagopus* " read " *vulgaris.*"
 144 „ 2 „ " *brachyrhyncus* " read " *brachyrhynchus.*"
 „ 14 „ " Halston " „ " Aston."
 146 „ 17 „ " boschas " „ " boscas."
 147 „ 21 Omit " B."
 „ 27 „ last line " A pair " to end.
 148 „ 11 for " 1888 " read " 1884."
 „ 16 Omit from " It nests " to " is known."
 155 „ 31 for " *C. Marylanda* " read " *Ortyx viginianus.*"
 158 „ 25 „ " *Ægialites* " „ " *Ægialitis.*"
 163 „ 9 Should read " because nets were stretched across to intercept the " cocks " as they shot by."
 164 „ 10 " Dunlin " should be in capitals.
 166 „ 6 Omit " B " and last 4 lines of paragraph.
 167 „ 25 for " *phæops* " read " *phæopus.*"
 170 „ 18 „ " The largest " read " One of the largest."
 173 „ 4 „ " 1882 " „ " 1884."
 174 „ 4 „ " *Podiceps* " „ " *Podicipes.*"
 190 „ Omit the entire article on the Sand Lizard. After examining specimens in the British Museum, I have come to the conclusion that in Shropshire only one species is found—the Common Lizard.
 194 „ 31 for " Mice, eggs, and young birds " read " Newts, and even Toads."
 198 „ 3 „ " Snake " read " Viper."
 199 „ 20 „ " *Colonella* " read " *Coronella.*"
 201 Omit lines 5 to 16, to " from lizards."
 212 line 7 Omit " and ears."
 „ 11 add " and ears."
 216 „ 21 for " known " read " know."
 Omit the entire article on the Edible Frog. Eyton was certainly mistaken as to the species.

FAUNA OF SHROPSHIRE.

CHAPTER I.

THE PHYSICAL FEATURES OF THE COUNTY.

HROPSHIRE, A THOUSAND YEARS AGO, presented an appearance so different to its aspect at the present day, that it is difficult to realize what would be the conditions of life at that period for both man and animals. We learn from various sources that the most marked feature of the County was the large area covered by continuous forests. Rev. T. Auden in a paper on "the Saxon Settlement of Shropshire," *(Transactions Caradoc and S. V. Field Club, Vol. I. p. 59)*, gives the names and extent of these various forests, and states that, broadly speaking, they covered two-thirds of the County, embracing the whole of South Shropshire, and extending some miles into the Northern part. " The County, as a whole, at the time of the Saxon invasion was therefore, as regards its natural features and position, not only remote, but much of it was uninviting and almost inaccessible." Mr. Thomas J. Davies in a paper on " the Severn " *(Op. cit. Vol. II. p. 14)*, says:—"The land bordering the Severn is cultivated and has definite boundaries, but, before the Roman invasion, and probably for centuries later, there were miles upon miles of forest, with swamps and vast

expanses of land intermittently covered with water . . . . .
the long graceful boat of to-day would have been a poor
friend to the Goidel, or Silurian and Brython of those times,
who would have been sorely at fault without the dug-out
canoe, or hide-covered coracle, which last is still used on the
Severn, Wye, and Dee." Mr. F. Rawdon Smith, in a letter
to the author says that the whole of the more elevated parts
of Shropshire were once densely wooded, though the only
remains we now have—beyond names and traditions—are
Wyre Forest, Shirlet, Kinlet, and the Forest of Mt St. Gilbert
(= Wrekin). He adds that the woods formed "one huge
nearly impassable timber belt, and that the lower ground
was swamp covered with alders, through which the Meese,
Sleap, Tern, Severn, etc., wandered; that this wilderness
was only opened at certain places and crossed by one or two
roads, and the whole district was *purposely* preserved in this
state, to help to keep out the turbulent Silures and, afterwards, the Welsh . . . . The major portion of
these enormous forests was cut down for fuel for forges,
etc., in the fifteenth and sixteenth centuries and afterwards." It is clear that so long as the forests remained
unreclaimed, and the marshy lands undrained, they would
afford shelter to birds and beasts, where they might live
secure from their greatest enemy—man. Indeed, the settle-
ments of men in Shropshire in early times were so scattered
and small that the whole population was probably not one
twentieth of the present number. As a consequence of this
state of things many birds and animals then flourished which
we now hardly think of as being Shropshire species at all,
such as the Wolf, Roe-buck, and Wild Boar; *possibly*, at a
yet earlier period, the Bear and the Beaver. The boom of
the Bittern would resound from the swamps, and the Kite

and Buzzard be seen daily sailing overhead in graceful gyrations. The Wild Cat might frequently be met with in the depths of the forest, while on the open land on its outskirts the noble Stag might be seen proudly raising his antlered head, and snuffing the breeze to detect the whereabouts of his enemy. The birds mentioned above are still with us, though in greatly diminished numbers : the mammals referred to have all vanished, and we may say, speaking broadly, that they disappeared with the forest which sheltered them. Perhaps it is hardly a matter for regret that the Wolf, Wild Cat, and Wild Boar are extinct here, but to the naturalist these animals, and the traces they have left behind them, are full of interest as connecting our limited fauna, not only with the distant past, but with the fauna of the Continent, and, through it, with that of the whole world. Turning now to SHROPSHIRE OF TO-DAY the first thing that strikes an observer is the variety of its surface. The whole area is only about 1,340 square miles, yet within this compass we have plateaus and plains, hills and vales, boggy flats and heathery moors, cornlands and pastures, wooded slopes and barren crags, meres and ponds, streams big and little, and —most important of all—the river Severn. The result of this combination of physical features (although some of them are necessarily on a small scale), is that the fauna likewise presents a varied aspect—that the fauna of Shropshire is as rich in species as that of any *inland* county in England. It is surprising what a number of sea and shore-birds come as visitors, not by any means always driven by storms, but in many cases attracted by such sheets of water as the meres at Ellesmere, etc., and by certain still reaches of the Severn, near Cressage and Melverley. Others again—Birds and Fishes—come to us by following the course of the Severn from the sea ; it is thus that such birds as Sandpipers reach us. The political boundaries of Shropshire are, of course, purely artificial, particularly in the north, and it is not necessary here to name the counties which adjoin, but there is one internal division so natural and well-marked that it may be well to study it with the aid of a map. This may be seen at a glance : the division of the County by the river Severn into two portions. If we examine the map closely we shall see that the two portions are of distinct characters. The part that lies to the S.W. is *hilly* or *wooded*, while that to the N.E. is comparatively *flat* and contains numerous *pools* and some tracts of *boggy moors*. It was in the first of these two portions that most of the ancient forests of Shropshire were situated (and there are still traces left, as in the Wyre Forest); and here it was that the animals before alluded to as living in these forests, were gradually driven to their last sanctuaries and exterminated before the advancing tide of civilization. The whole district is more or less mountainous. The highest hill is Brown Clee (1792 feet), while considerable tracts of land—forming parts of the Longmynd range, with the adjoining hills—are over 1,000 feet in elevation. The following are some of the principal heights :

Titterstone Clee	... 1749	Caradoc	... 1506
Stiperstones	... 1731	Cothercot	... 1457
Longmynd	... 1696	Wrekin	... 1335
*Corndon	... 1684	*Breidden	... 1202
Clun Forest	... 1619	Wenlock Edge..	900

Almost all these high lands belong to the older geological formations; they consist of hard rocks, and are barren or of a heathy character. The open moorland on the Longmynd is the resort of many kinds of birds which are seldom met with

* On the border of Shropshire, in Montgomeryshire.

elsewhere in Shropshire. Ravens used to reside there, and here we still find such birds as the Twite, Snow Bunting, Grouse, and Curlew, whilst the intervening wooded valleys teem with a variety of birds and animals that love sheltered places. Across the Severn the country is very different. With the exception of the isolated Wrekin (which may be looked upon as an outlier of the range of hills opposite to it), there is no hill over 1,000 feet high, the land being open and most of it cultivated. There are no extensive woods, and the country is a plain, overlying the New Red Sandstone, which is seen forming low hills at Hawkstone (420 feet), Grinshill (629 ft.) and Nesscliffe (500 ft.) At Whixall, Baggy Moor, and the Weald Moors, are peaty or wet bogs, the haunt of wading birds and amphibians, whilst at Ellesmere, Whitchurch, and around Baschurch and Berrington, are many meres and pools attractive to water-fowls. Before concluding this chaper it may be well to give an account of the course of the river Severn through the County. It enters Shropshire on the N.W. beneath the shadow of the grand old Breidden Hills. It is not a large river there, but, just as it crosses the boundary, it receives on its left bank the Vyrnwy, and thenceforward rolls on with a considerable volume of water. The next part of its course is through flat alluvial deposits ; the slope is so slight that the current is sluggish, and the river makes so many bends that between Pool Quay and Shrewsbury—less than twenty miles apart—its course measures forty-two miles. Near Shrewsbury is to be seen a place where, at some distant time, the Severn changed its course : the present river-bed forms a horse-shoe bend nearly surrounding the town ; the old course ran directly across the isthmus. Below Shrewsbury the river runs, still circuitously, with alternate quiet

reaches and rapid fords, as far as Buildwas.   There it enters
a narrow defile, and flows with a deep strong current,
between well-defined banks, and in a fairly straight course,
southwards to Bewdley, where it quits Shropshire.   Up to
the early part of the present century the river was navigable
for barges as far up as Pool Quay, and there were far more
fish (especially Salmon), in the upper waters, than at present.
The Severn is very liable to floods, whenever there
are heavy rains on the Montgomeryshire hills; particularly
since improved land-drainage has caused rain-water to find
its way more speedily into the river.   At such times the
water is turbid from the quantities of fine mud washed into
it from the surface of the plain.   The Vyrnwy is never
muddy because it flows over a rocky bed throughout its
course; when flooded it becomes coloured a clear brown
from the quantity of water washed into it from the peat-bogs
around its source.   Even at Shrewsbury it is possible to
tell, by the colour of the water, whether the Severn, or
Vyrnwy, or both, are in flood.   There are several smaller
rivers in the County; nearly all tributaries of the Severn.
Yes, we may well invoke a blessing upon "Sabrina fair,"
and, considering all that she has done to add to the interest
and beauty of one of the most interesting and beautiful
counties of England, say, in the language of Milton :—

> " May thy brimmèd waves for this
>   Their full tribute never miss
> From a thousand petty rills,
>   That tumble down the snowy hills.
> Summer drouth, or singèd air
>   Never scorch thy tresses fair.
> May thy lofty head be crown'd
>   With many a tow'r and terrace round,
> And here and there thy banks upon
>   With groves of myrrh and cinnamon."   *Comus.*

JOHN ROCKE.
1817—1881.

JOHN SHAW.
1816—1888.

WILLIAM PINCHES.
1802—1849.

Photos by R L. Bartlett Shrewsbury.

THOS. BODENHAM.
1804 1873.

HENRY SHAW.
1812---1887.

WM. FRANKLIN.
1814— 1885.

REV. WM. HOUGHTON.
1828—1895.

CHAPTER II.

SHROPSHIRE NATURALISTS.

A NYONE writing a book on the Natural History of Shropshire must, necessarily, avail himself of the work done by his predecessors in the same field. The County of Salop may well be proud of the noble army of her sons who have distinguished themselves as naturalists. Charles Darwin stands in the forefront as the man who fought for a great principle and infused a new spirit into the "dry bones" of science; but, it is not with such as he that we are concerned in the present volume; no—the book aims only at giving a complete account of the Vertebrate Animals of the County, and therefore owes much to those patient workers who during the last 80 years have sought after, and recorded, all the interesting occurrences in their own neighbourhood, and so preserved to us a heritage of valuable facts that would otherwise have been lost. The first list was published by T. C. EYTON in 1838, in the "Annals and Magazine of Natural History." It is entitled "An attempt to ascertain the Fauna of Shropshire and North Wales," and is the only one that records the Mammals, Reptiles, and Fishes, as well as the Birds. The next list was written by JOHN ROCKE, and published in the *Zoologist* for 1864-5. This gives an excellent account of the Birds of Shropshire, with ample details up to that date. Mr. Rocke not only wrote about the birds, but made a splendid collection of specimens at his house at Clungunford. A third list of Birds was prepared by W. E. BECKWITH, and published in the

" Transactions of the Shropshire Archæological Society " in
1879. In this list are many additional records of rare Birds
as well as species not recorded previously. Later on (1887),
Mr. Beckwith commenced a series of papers in the same
publication, entitled " Notes on Shropshire Birds." These
are of the greatest value and interest, as they give full details
of the habits of all our commoner Wild Birds ; drawn mainly
from his own personal observation. These notes would have
been published as a book, but his death in 1892, unfortunately, put an end to them when only half completed.* In
1897 Mr. G. H. Paddock, of Newport, Salop, published,
privately, a book on " Shropshire Birds," containing notes on
their habits from his own observations, and added two
species to the list, as well as recording the nesting of several
Birds in the County not previously noticed. Besides the
above who have written on the subject, our County owes
much to the two taxidermists—WILLIAM FRANKLIN
and HARRY SHAW—whose skill and originality were so
universally recognised that Shrewsbury became the " Mecca
of Ornithologists," and their services were sought by
collectors and sportsmen from all parts of the kingdom.
Much, too, is owing to WILLIAM PINCHES, THOMAS
BODENHAM, and LORD HILL, who made collections of Birds
in which Shropshire specimens occupy a prominent place.

William Edmund Beckwith was born at Eaton
Constantine, Dec. 17, 1844. He was educated at Bridgnorth, and afterwards at King's College, London. He
then came to reside at Eaton Constantine, and lived
there till the death of his father (Rector of that place) in
1888. From his earliest years he showed his strong love

---

*NOTE.—If this book is successful, possibly arrangements may be made to
complete these notes and publish them as a supplement.

of natural history, and in after life was most fortunate in having a great deal of leisure time, which he spent in the pursuit of his favourite study. He will always be best known for his accomplishments as an ornithologist, but his knowledge of other branches of natural science, especially botany, was extensive. Birds he studied all his life, in season and out of season, in the cabinet and at large, on land and water, by night and by day; and he thus gradually became known all over the County as an authority on the subject, and all who happened to come across any rare bird, or remarkable occurrence, would write and tell him about it. He was exceedingly careful in recording such things, and would travel any distance to verify a doubtful observation or to visit one of his correspondents. He made copies of anything that struck him when reading books on Birds, etc., and filled many scrap-books with cuttings from the *Field* and similar publications. He frequently contributed short notes to the columns of that paper, and wrote in it a highly interesting article on the Sea Birds that visit Shropshire. His important writings on Shropshire Birds have already been mentioned. It is much to be regretted that these never appeared as a book, with the numerous other particulars which he would, doubtless, have added from his inexhaustible stores of information. The author also regrets that he is unable to give any portrait of Mr. Beckwith, for he never would sit for his photograph, and there is not even a sketch of him in existence. He was a very big man, and of a florid complexion, with brown hair, and light coloured eyes. He was so enthusiastic about his favourite pursuit that all who had similar tastes courted his society, and were charmed by his conversation. In observing Birds out of doors he always used a good field-glass, and recommended others to do the same. Although he never made a systematic collection of Birds, he accumulated during his life a large number of the rarer species, and these are still in the possession of his family at Radbrook. One of his most recent acquisitions, and one of which he was very proud, was the "White's Thrush" shot at Moreton Corbet in 1892, a photograph of which is given in this book. He lived at Eaton Constantine till the year 1889, when he moved to Radbrook House, near Shrewsbury, and here, to the great grief of all who knew him, he died on July 22nd, 1892. He was buried at Eaton Constantine.

Thomas Bodenham, born 1804. Very little is known about this gentleman except that he had independent means. He lived first at Hook-a-gate and seems to have been a good botanist, for he contributed many notes on the plants found in that neighbourhood to Leighton's "Flora of Shropshire." He afterwards came to reside in Column Buildings, Shrewsbury, and subsequently at a larger house called Sunfield. He was of a very retiring disposition, and rarely left home, especially during the later years of his life, when he was a chronic invalid. From 1849 till his death he was a member of the Swan Hill Congregational Church and a liberal donor to its funds. It is not known how or when he took up the study of ornithology, but it was probably late in life, after he moved to Sunfield. He then made the acquaintance of Mr. Beckwith, who used to visit him so frequently that a room was kept always ready for him, and was called "Mr. Beckwith's room." Mr. Bodenham seems to have devoted himself especially

to making a collection of Shropshire Birds, and in many
instances the only specimens of particular species ever
obtained in the County found their way into his hands.
In both Beckwith's and Rocke's lists we frequently
meet with the words " in Mr. Bodenham's collection,"
yet now the collection no longer exists. At his death
all his possessions were sold and the property divided
amongst various relations. He was buried at Pulverbatch, December 18th, 1873.

Thomas Campbell Eyton, was born at Eyton Hall, near
Wellington, Salop, September 10, 1809. He took up
the study of natural history at an early age, and
numbered amongst his friends Agassiz, Asa Gray,
Charles Darwin, A. R. Wallace, and Professor Owen. In
1836 he published a " History of the Rarer British
Birds," with beautiful wood-cut illustrations, the work
of a local engraver named Marks. In 1838 appeared his
" Monograph of the Anatidæ, or Duck Tribe." A small
reprint of this book appeared in 1869. The same
year he published in the *Annals and Magazine of Natural
History* (a periodical then just started) a highly interesting paper, entitled " An attempt to ascertain the
Fauna of Shropshire and North Wales." This list
gives the Mammals, Birds, Reptiles, and Fishes of the
district, and forms the foundation on which subsequent
naturalists have worked. The reason that Shropshire is
here associated with North Wales was that the paper
was written by Eyton for a Society formed a few years
previously (1835) in which he took a strong personal
interest—the Shropshire and North Wales Natural
History and Antiquarian Society. He wished the paper
to cover the same area as that of the sphere of the

Society's operations.   The old Museum on College Hill
was formed by the same Society, though later on
transferred to the Corporation of Shrewsbury.   After
coming into the estate in 1855, Eyton built a large
museum at Eyton Hall, and amassed here a marvellous
collection of skins and skeletons of Birds from all parts
of the world.   Most of the skeletons were prepared and
mounted by his own hands, and he issued in 1858 a
catalogue of these, a copy of which is in the Reference
Library at Shrewsbury.   Later on (1871-8) he published
a large work, entitled " Osteologia Avium," illustrated
by drawings made from the skeletons in his museum.
He also wrote several other works on various subjects.
An ardent naturalist and sportsman, Eyton was always
most ready to help any who took an interest in his
pursuits, and visitors came from far and near to see him
and his collections.   His friendship for Darwin continued
to the end of his life, though he was strongly opposed
to his theory of Natural Selection, and was much vexed
at finding some of his own observations on the habits of
pigeons used by Darwin in support of that hypothesis.
He died October 25, 1880, and his collections were then
dispersed, but the shells were presented to the Shrewsbury Museum.

William Franklin was born at Serne Abbot, Dorsetshire,
in 1814.   At an early age he was brought by his father,
a hatter, to Shrewsbury, and there started in business as
a hairdresser.   His fondness for wild Birds and Animals
led to the wish to preserve them, and a travelling
journeyman gave him his first and only lesson in the
mystery of skinning, preserving, and setting up specimens.
It will be easily understood that his difficulties were

many, but his intense love of the work carried him on to success, and he ultimately became so skilful that he was able to relinquish his original trade and take to taxidermy as a profession. It is mainly to him that we are indebted for the complete revolution wrought about this time in the method of mounting Birds. The old method was to set Birds up all alike on stiff wooden perches with a label bearing the name and other particulars. Mr. Franklin made his cases real works of art. He studied the habits of the various species and imitated their attitudes to perfection, while the accessories were always in keeping with the natural habitat of the Bird. As an instance we need only refer to the life-like appearance of the pair of Ospreys pourtrayed in the frontispiece to this volume. His best claim to fame rests, however, upon his invention of an entirely new method of making artificial rockwork on which to place his Birds. The old plan was to make a foundation of wood, cover it with paper or canvas, coat this with glue, and then—while still wet—sprinkle it over with sand; the result was quite unlike anything in nature! Mr. Franklin sought to improve on this, and, after many experiments, finally decided on peat as the basis of his rock-work, coating it with paint which his innate sense of colour enabled him to use with most artistic and natural effects: this plan with various modifications has since become general, though probably few of the men who afterwards adopted his method had any idea who invented it, for the originator was of a most quiet and unassuming disposition, and shrank from publicity in any shape. He died in 1885 after a lingering and painful illness, and, to the last, retained his intense love of animated nature.

Rowland Hill, second Viscount, of Hawkstone (nephew of
the "Great Lord Hill"), was born in 1800, and succeeded
to the title and estates in 1842.   He had represented
Shropshire in Parliament during three sessions, and was
a member at the time of his succession to the peerage.
In his public life he was exceedingly popular, his kindly
manner and high intellectual qualities fitting him in an
eminent degree for such a position as that of Lord-
lieutenant of the County, which he occupied subsequently.   He was an ardent supporter of the Salop
Infirmary and other charitable institutions, and profuse
in his liberality.   Always fond of sport of all kinds, he
developed a taste for natural history, and very soon after
coming into the estate he determined to form a private
museum, which should be as complete and beautiful as
money and assiduous research could make it.   To
accommodate it he built a new wing to the existing
mansion, lofty and well lighted, and commissioned
Henry and John Shaw to fit up a series of cases
appropriate to the various classes of Birds and Mammals.
In the short space of five years there was gathered
together a complete series of all the then known species
of British Birds and Mammals, such as was never before
seen in any private collection.   Here also most of the
Shropshire specimens of rare local occurrence found a
fitting home, and these are mentioned frequently in the
following pages.   Not content with studying and collecting our native Animals, Lord Hill experimented in the
Park with the acclimatization of such foreign species as
the noble African Antelope, the Eland; and the curious
Australian Bird, the Emu.   One of his sons had a
narrow escape from an infuriated Eland, while now and

then an Emu would escape from the paddock and only be captured after an exciting chase of many miles. In his natural history pursuits Lord Hill was aided by his eldest son, Rowland (afterwards third Viscount), who personally not only collected and obtained Birds for the museum, both in England and abroad, but stuffed many of them under the supervision of Henry Shaw. In conjunction with Sir Thomas Boughey, the 3rd Viscount started the celebrated pack of Otter Hounds, but as he was also Master of the Fox Hounds, and could not hunt both, he handed over the former to his brother Geoffrey, who thenceforward became known far and near for the enthusiasm and skill he displayed in otter-hunting. The portrait of Lord Hill in this volume is an excellent likeness taken when he was about sixty-five. He died January 2, 1875, full of years and honours, and beloved by tenants and friends, who erected in Hawkstone Park a beautiful marble statue to his memory.

Rev. William Houghton was born in 1828, and was for many years Rector of Preston-on-the-Weald-Moors, near Wellington, Salop. His taste for natural history was very comprehensive, ranging from the higher vertebrates to microscopic organisms, and he wrote several works on the natural history of the ancients. He was elected a Fellow of the Linnean Society in January, 1859, and was an active member of the Severn Valley Field Club, to whom he read a most interesting paper on "The natural history of the Weald-Moors." Amongst the books which he wrote upon popular natural history were " Country Walks of a Naturalist " (no date); " Seaside Walks of a Naturalist," 1880; " Sketches of British Insects," 1877; and a large folio in two volumes on " British

Fresh-water Fishes," 1879, illustrated by splendid chromo-lithographic plates of every species. With the exception of the paper on the " Weald-Moors," he did not write anything of special local interest, but he was a good observer and noted any rare or unusual occurrence such as we find recorded in this volume under the heads of "Dotterel" and "Black-tailed Godwit." The portrait given here is copied from one in the Shrewsbury Museum, and represents him in the prime of life, when he was a man of fine physique. His health failed very much latterly, and he had to relinquish his clerical duties and retire to Tenby, where he died September 3, 1895. During the last few years of his life he was in receipt of a government pension, granted in recognition of the value of his writings. Copies of most of his works are to be found in the Reference Library at Shrewsbury.

William Pinches, of Ticklerton Court, near Church Stretton, was born 1802, and died, unmarried, in 1849. The family of this name had resided on their properties at Ticklerton and Harton for generations. He was a fine, big man, with a florid complexion, and was a keen lover of sport of all kinds. The old kennels are still standing at Ticklerton in which he kept the greater part of the " United " Pack of Fox Hounds ; in those days, a rough-haired breed of dogs. He was also a good naturalist, and had a fine library of books on sport and natural history, as well as a large collection of Birds, many of them very rare locally, and in great part obtained by his own gun. One of the gems of the collection was the Great Auk, perhaps the finest specimen in existence on account of the perfect state of its plumage. It was sold

and passed into the possession of Sir William Milner.
It is remarkable that although Mr. Pinches' collection
must have been in existence and contained many Birds
of great local interest—no allusion to it appears in the
lists of either Mr. Eyton or Mr. Rocke. Several of these
specimens are mentioned in the present volume. Most
of the collection was made between 1830 and 1840, and
to render the series more complete Mr. Pinches procured, through Mr. Shaw and others, imported skins of
such Birds as he was unable to find in the neighbourhood. The rarest local Bird is the Squacco Heron—the
only one ever obtained in Shropshire. About the year
1840, Mr. Pinches, with his brother-in-law, Mr. Buddicom, procured some Red Grouse from Yorkshire, and
turned them out on the Longmynd, of which they had
the shooting. It is said that these were the first Red
Grouse ever known on the hills (though there were plenty
of Black Game), and that from them have sprung the large
numbers now found there. The portrait here given is from
a pencil sketch in the possession of Mr. Buddicom, and
shows Mr. Pinches in his hunting cap. The collection
of Birds is still in the same house, and passed with
the other property to Mr. Buddicom, father of the
present owner.

John Rocke, eldest son of the Rev. John Rocke, was
born at the old Rectory, Clungunford, June 10, 1817,
He went to school at Bath, and thence to Harrow,
where his career as a naturalist commenced, for, whilst
there, he began to collect and stuff Birds, thus forming
a nucleus of the museum which became the delight and
pride of the later years of his life. In 1836 he
entered at Trinity College, Cambridge, and took his

B.A. degree in 1839. In 1841 he and his friend Mr.
Solly went a fishing expedition to Norway—a country
then almost unknown to Englishmen. This visit was
repeated the following year. On the death of his father
in 1849 he succeeded to Clungunford Hall and estate,
and at once began to add largely to his collection of
British Birds. In pursuance of this object he cultivated
the acquaintance of the leading ornithologists, and
numbered amongst his most intimate friends Mr. Gould,
Mr. Howard Saunders, Mr. Hancock (of Newcastle-on-
Tyne), and many others of that ilk, gaining much
valuable information from his intercourse with them, and
meeting them periodically at their houses, or rooms in
Tenterden Street. His Birds, however, still remained in
their small cases. In July, 1860, the museum at Clungunford was begun, and the one side with its large cases
completed in 1861. The Birds, which had greatly
increased in number, were classified and mounted by
Henry Shaw. Mr. Rocke's friends in Tenterden Street
sent down numberless skins for selection, and the Great
Bustard—a most beautiful specimen—was offered by Mr.
Howard Saunders, with the remark, " Your museum is
the only one in the world worthy of it." The purchase
of the Great Auk and its egg rendered this one of the
most perfect private collections of British Birds in
existence. Every British Bird is here represented,
both male and female, and in many cases the young
also are added. Every specimen is well set up, and is
in perfect plumage ; the case of Falcons is unique. No
expense or trouble was spared to render the museum
perfect in every respect. To within a very short time of
his death Mr. Rocke was at work at his favourite study.

Nor was Oology forgotten :—He collected and arranged the egg of every British Bird, including many remarkable varieties. In 1864 and 1865 Mr. Rocke published his "Notes on the Birds of Shropshire" in the *Zoologist*. In 1878 eyesight and health began to fail, but he continued to manifest the keenest interest in his museum, and had begun a small collection of the young of many of the Birds in it which up to that time he had been unable to procure. This, however, he never completed. He died April 3, 1881, at the age of 64. Mr. Rocke was J.P. and D.L. for the County of Salop, was Sheriff in 1869, and was for many years Lieutenant in the South Salopian Yeomanry Cavalry.

Henry Shaw, familiarly known as "Harry" Shaw, was born at Tarporley, Cheshire, October 3, 1812. He came to Shrewsbury as a boy, and was educated by Mr. David Parkes, of Castle Street, a well-known man in his day. He must have been a boy of quick perceptions, for in after life he displayed an amount of information on his particular line of work which few men could equal. His father's shop was a small one in Shoplatch, and was demolished in 1868 to make room for the New Market Hall. Henry and his brother John worked at taxidermy under their father, and for some time after his death remained in partnership but eventually they separated. Both started in business at Shrewsbury and were clever men at their profession, but Harry, owing to his more genial nature, got on much better than John with the country gentlemen, and thus secured most of them as his patrons. He secured the orders to mount and arrange the collection of Birds at Hawkstone, Clungunford, and Ludlow Museum, and received

large sums of money for the work done at those places.
Lord Hill appointed him curator to his collection at an
annual salary, and by this means Henry Shaw became
so famous as a taxidermist that orders flowed in to him
from all parts of the kingdom. The first "Short-toed
Lark" found in Britain was recognised by him and sent
up to Mr. Yarrell, who recorded it in his "History of
British Birds."   He moved into his well-known shop at
45 High Street in 1870. Besides the collections named
above, Henry Shaw was mainly instrumental in arrang-
ing and mounting those belonging to Col. Wingfield, at
Onslow ; Earl Powis, at Powis Castle ; Mr. Naylor, at
Leighton Hall; the Duke of Westminster, at Eaton
Hall; the Duke of Portland, Welbeck Abbey; and
many others. This work at Welbeck was the last that
he lived to accomplish, for he died after only a few days'
illness October 7th, 1887, a hale, strong man, at 75; he
was a Hercules as a young man, and many tales are told
of his prowess as a fighter. Probably few other men
could say that they had had through their hands *three*
specimens of the rare, and now extinct, Bird—the Great
Auk. His love of Sport continued to the end of his
life ; an expert salmon fisher, he rented part of the Wye
and caught a rare lot of fish there annually ; indeed, he
probably owed his death to exposure to cold while
salmon fishing at Builth, which brought on pleurisy.
His snow-white hair, big, powerful figure, and ruddy
countenance must yet linger in the memory of most
residents in Shrewsbury.

John Shaw was a man of shorter stature and less robust
frame. On the dissolution of the partnership with his
brother he took a shop near the High Street end of the

Wyle Cop, opposite St. Julian's Church, ultimately removing to premises three doors away, now No. 82, of which he became the owner, and which he occupied until his death. Although not called away so much as his brother Henry, he had a large business connection throughout, and far beyond, the limits of the County. He was a man with characteristics peculiarly his own. Confident of his own ability, he liked his customers to have confidence in him, and this, it generally turned out, was not misplaced In his later years he might invariably be met taking his constitutional walk in the Quarry between nine and ten o'clock in the morning. By those—and they were many—who could see through his superficial manners, it was found that they formed but a thin veneer concealing a thoroughly genial and kindly disposition. He was an art critic of no mean ability, and more than one local artist owed his rise to fame to the discriminating taste of John Shaw. He never married, but his conduct in private life engendered a love for him and a grief at his loss known to but few outside his immediate circle. His death took place in July 1888, at the age of seventy-two.

CHAPTER III.

MAMMALS.

IN considering the animals of an inland County like Shropshire, several points are of importance as affecting the
whole fauna. In the first place, the County is only an
arbitrarily defined portion of a country, and that country an
island. The insular position of Great Britain renders it
almost impossible for any creatures to reach it from outside
unless they can either fly in the air or swim in the sea. Now
most Mammals are terrestrial, but one group (the Bats) can
fly like birds, while another group (Whales and Seals) can
swim. The Bats, however, unlike many birds, do not
migrate; instead of that they hibernate or sleep through the
winter. They usually live and seek their food and die in the
same district where they were born. The Whales and Seals
roam afar in the sea, but rarely ascend rivers far enough to
reach so inland a County. Thus it happens that the
number of species of Mammals is very limited, and cannot
be added to from outside unless by human agency, and any
species, once extinct, cannot occur again. For this very
reason our Mammals are particularly interesting to the
naturalist, and proportionately more space is devoted to
them in this volume than to the Birds. This course seems
the more necessary because—few as they are—the number of
species is gradually but surely lessening. The Wolf, the
Roebuck, and the Wild Boar have been long extinct; the
Black Rat and the Pine Martin disappeared this century;
while the Polecat is on the verge of extinction. It was the

Photo by R. L. Bartlett, Shrewsbury.

T. C. EYTON.

1809—1880.

Photo by R. L. Bartlett. Shrewsbury.

LORD HILL.

consideration of the necessity of collecting and recording data about these creatures before they became utterly lost that led the writer to prepare notes for this volume, though he did not at first contemplate undertaking a book on the whole of our vertebrates. While no attempt has been made to give fossil forms, several Mammals are included that have become extinct since the Roman invasion. A noticeable characteristic of our Mammals is that they are, almost without exception, nocturnal in their habits, possibly from fear of man ; they conceal themselves by day in holes or amongst herbage ; even those which do come abroad by day fly at his approach. This makes the study of their habits in a state of nature one of peculiar difficulty, for there are few naturalists who will spend their nights out of doors for the sake of learning some new fact concerning them. A great deal, however, may be learnt by going to some secluded wood on a fine summer evening and sitting perfectly still. Rabbits, Stoats, various Mice, Squirrels, etc., will soon appear and disport themselves within a few feet of the observer apparently unconscious of his presence, but—if he stir ever so lightly—they instantly vanish into obscurity ! After a short interval, if he remain quiet, they re-appear from their hiding places and renew their occupations. Birds, too, treated in the same manner will allow themselves to be studied at much closer quarters than in any other way, and that in daylight as well as by night. The late Mr. Beckwith used to say that in studying Birds out of doors a good field-glass was a much more useful weapon than a gun ! All practical field naturalists will endorse this opinion. The nomenclature adopted in this book is that of Lydekker's " Handbook to the British Mammalia," an excellent manual to which the author is indebted for many details in the following pages.

LONG-EARED BAT. Bats are generally regarded by
Plecotus auritus. country folk with feelings of super-
stitious dread or aversion, perhaps
because they are creatures of the night. They are, however, not only harmless but very useful animals, as they
perform, by night, the work done by the swallows, etc.
during the day—catching insects on the wing. There
is no doubt that were they not kept in check by bats,
many of the smaller moths would multiply unduly and
their larvæ, or caterpillars, do an immense amount of
damage. It is true the caterpillars are eaten in large
numbers by various birds, but the adult insects are only
abroad during the night when there are few birds to
prey on them. It is then that the bats come forth to
" act the policeman." Of the eleven species of Bat
recognised as British, only six are known to occur in
Shropshire, and, of these, the Long-eared Bat is the
most numerous. The very large ears—as long as the
body—are so remarkable, that no one can possibly mistake this for any other kind of bat. It is very gregarious,
and during the day goes to roost in groups of a dozen or
more in crevices in rocks, old buildings, bridges, in the
roofs of houses, and sometimes in holes in trees. It
invariably sleeps suspended by its hind toes, head downwards. About dusk the bats may be seen leaving their
hiding places in a long procession, and they return in the
same way about dawn. The Winter sleep of this Bat
is a long one, generally lasting without interruption
from early Autumn to late in the Spring. When at rest the
long ears are doubled backwards and laid along the sides
of the head and shoulders, and the small-pointed tragus,
or inner ear, stands out so conspicuously that it looks just

like the real ear. If kept in confinement this Bat will
eat bits of raw meat, and it is most interesting to watch
its motions. While eating it brings the wings forward
to cover the head. When about to fly it hangs head
downward, bends the head upward, unfolds the large
ears, then expands the wings and darts off. Its flight is
strong and rapid, and it can turn in the air with agility.
The ears are always expanded while on the wing, and
the membranes of both ears and wings are so sensitive
that Bats have been supposed to possess a sixth sense
enabling them to perceive the smallest object in the air
ahead of them. Their sense of smell is said to be very
acute, but it is unlikely that they are keen sighted, as
the eyes are small. Although Bats are able to fly like
Birds, the structure of the wings is very different.
There are no feathers, but a thin flexible membrane
consisting of two layers of skin stretched out upon the
three elongated joints of the four fingers and continued
along the arm and sides of the body to the hind limbs and
tail. The thumb is short and terminates in a sharp curved
claw, but the fingers are clawless. The hind limbs are
short, and the knees directed outwards instead of forwards.
To fold the wings the fingers are brought close together
and laid backwards along the side of the body, and in
this position the thumb with its claw projects forward
below the head. If the Bat is on the ground it stands
on its thumbs and hind feet, and its method of walking
is by a kind of shuffle, dragging forward first one side
and then the other by fixing the thumb-claw into any
roughness of the surface over which it is travelling. It
moves more easily this way upwards than on a horizontal
surface, but when it reaches a place where it wishes to

rest, almost always turns round and suspends itself by
its hind feet, which, like the thumbs, are furnished with
sharp hooked claws.   The Long-eared Bat while on the
wing utters a shrill squeak: so shrill or high-pitched is
this cry that many people's ears are unable to perceive
it.   The female is said to produce only one at a birth,
and folds her wings around it while suckling, the baby
also clinging to the mother by its little claws.   Length
from head to tail, $3\frac{1}{4}$ inches.   Colour, brown above,
whitish beneath, but the basal part of each hair is
black.

NOCTULE, or GREAT BAT.   This is the largest of our
Vesperugo noctula.   Bats, and when seen in the air appears
                                even larger than it is on account of the
width of the wing membrane and the light colour of its
fur.   It is partial to the neighbourhood of trees, and
usually sleeps in holes in them.   It comes abroad early
in the evening and flies high in the air where the large
beetles, on which it loves to feed, chiefly abound.   This
habit of flying high up in the air is as characteristic of
this species amongst Bats as it is of the Swift amongst
Birds.   It is gregarious, as many as fifty having been
taken out of one hole, and has a peculiar disagreeable
odour, which is very perceptible when any number are
found together.   The two sexes are said to hibernate
in distinct colonies from October to April, and are found
then in larger numbers together than during summer.
One, rarely two, young are produced at a birth, and they
are at first blind and naked.   The Noctule seems to be
pretty generally distributed through Shropshire, especially
in the neighbourhood of water, but is not so numerous
as the Long-eared.   Length from head to tail, 5 inches;
Colour, light yellowish brown, with blackish wings.

PIPISTRELLE, or COMMON BAT. Although this Bat
V. pipistrellus. is common all over the County, it is
not so numerous in Shropshire as the
Long-eared. It is, however, more often seen than any
other species because it comes abroad at times when all
the other Bats are hibernating. Indeed, even in mid-
winter the Pipistrelle has been known to come out,
roused from its slumbers by unusually warm weather,
and it commonly terminates its period of hibernation in
March. It is found in all kinds of holes and corners in
buildings, and numbers may be seen on any warm
evening flying over the English and Welsh Bridges at
Shrewsbury, and doubtless they spend the day in the
crannies below the arches of the same. It is very likely,
however, that some of the Bats seen there belong to the
next species, which is especially fond of the neighbourhood of water. The Pipistrelle is the smallest of our
Bats, and flies with an unsteady fluttering motion of the
wings—whence country people call it the flitter-mouse—
turning and winding in all directions. It never flies
high up like the Noctule, nor so low down as the
Daubenton's Bat, but generally at a moderate elevation.
It does not avoid man, but may often be seen flitting
along streets and around houses even in large towns.
It even comes out occasionally in daylight, and the
writer once saw one in Scotland hawking for flies along
with numbers of Swallows and Martins in the full blaze
of a July *mid-day* sun, and apparently quite as happy
as the Birds. Like the other Bats, the Pipistrelle has
usually only a single young one at a birth. Length from
head to tail, 3 inches. Colour, dark brown. The ears
are small, and the face deeply depressed behind the black

muzzle, giving the Bat a peculiarly wizened appearance.
DAUBENTON'S BAT is best known by its extreme fond-
Vespertilio Daubentoni. ness for water. It comes out from its
lurking places in trees or river banks
late in the evening and skims along just above the surface
of the water, much in the manner of a Swallow, though
it rarely touches it with its wings. Mr. J. Steele Elliott
reports that this Bat is common at the mouth of Dowles
Brook, in the Wyre Forest, and along the river Severn
generally throughout the County. It is very probable
that it occurs in suitable situations in other parts of
the County, but, no doubt, the late hour at which it
comes out, causes it to be overlooked, and makes it
difficult to procure specimens. A peculiar aspect is
imparted to the face of Daubenton's Bat by the cheeks
being swelled out into glandular protuberances. The
ears are rounded and rather small. Length from head
to tail, 3⅝ inches. Colour, brown above and dirty white
beneath, the basal half of each hair being dark brown.

NATTERER'S, or REDDISH-GREY BAT. In the
V. Nattereri. additional notes to his "Fauna of
Shropshire and N. Wales" published
in the *Annals and Mag. of Nat. History* vol. IV., p. 396,
Eyton states that he had a specimen of this Bat taken
at Eyton, near Wellington, Salop. This was in 1840,
and is still the only record for the County, though as the
species is not a rare one it almost certainly occurs here,
but passes unrecognised. This Bat is distinguished by
the very long and thick fur; by the lightness of colour,
especially of the under-parts; and by the fringed inter-
femoral membrane. In its general habits it resembles
the Pipistrelle but seems to frequent buildings almost

exclusively, especially roofs of churches, and is sometimes found in large numbers in such situations. The general colour is light red above and whitish beneath, but the basal part of each hair is dark brown. Total length 3⅝ inches, of which the tail measures nearly half.

WHISKERED BAT. Mr. Chas. Oldham, of Sale, near
V. mystacinus. Manchester writes in the *Zoologist* 1890, p. 349, "In the beginning of June of this year, about six p.m. one [of these Bats] was knocked down with a stick in a garden at Hanwood, near Shrewsbury." This is the only record for Shropshire. According to Mr. Lyddekker "Essentially a solitary species, although occasionally seen in small companies attracted by an abundance of food, the Whiskered Bat appears to frequent for the purposes of hibernation, either hollow trees, the roofs of buildings, or caverns. It makes its appearance early in the evening, and flies swiftly in a manner very similar to the Pipistrelle. It often exhibits a preference for the neighbourhood of water, over the surface of which it skims." It is distinguished from other Bats by the wing membrane starting from the base of the outer toe and by the presence of long hairs on the face covering the upper lip—hence its name. Colour reddish-brown above and greyish beneath. Total length nearly 3 inches.

HEDGEHOG, or **Urchin.** The Hedgehog, with his
Erinaceus europæus. armour of sharp spines, is so different from all other British animals that it is easily recognised by the least scientific observer, and its method of defending itself, by coiling up into a ball bristling with points, is familiar to all. Yet many curious errors have passed current regarding its nature and habits. For instance, it has been said to climb trees and

carry off the fruit impaled on its spines; an idea that
probably originated from its accidentally falling on an
apple while coiled up, and so going off with it adhering
to its back, for the Hedgehog can drop uninjured from a
considerable height, the elasticity of the spines breaking
its fall. It is certain that it does not eat fruit, or,
indeed, vegetable matter of any kind, its food consisting
for the most part of worms, slugs, beetles, and young
birds; they are also fond of birds' eggs, and in short will
eat any small animal, bird, or reptile that they can
manage to kill. They sometimes eat poultry, attacking
them at night when asleep on the nest, and putting them
to a horrible lingering death by eating into the soft parts
of their bodies. The writer has had two hens killed in
this way and could not imagine what animal had done
the mischief, till he read in the *Field*, an account of a
similar case in which a Hedgehog was the culprit. It
is stated on good authority that they will kill and eat
snakes and vipers, approaching them with caution, biting
them on the neck, and then instantly coiling up. If it is
a viper that is thus attacked, it turns on the Hedgehog
to bite it (one bite would be fatal), but only succeeds
in lacerating itself against the sharp spines, and soon
succumbs to its injuries. In captivity the Hedgehog is
useful in destroying cockroaches in kitchens, and the
writer had one that used to come regularly every night
for a saucerful of bread-and-milk. Hedgehogs generally
swarm with fleas of a species peculiar to themselves.
They rarely come abroad till late at night, but may
then be seen running eagerly about poking their sharp
snouts under leaves and into crannies in search of worms,
etc., and occasionally giving utterance to short grunts

like little pigs. They hibernate through the winter
months buried amongst masses of dead leaves or
under the roots of large trees. They must, necessarily,
sleep continuously through the winter for—as their
food is entirely of an animal nature—they cannot
lay up stores, and nothing can then be obtained
from the frozen earth. The female has six teats,
and usually has about four young at a birth, though
sometimes as many as seven. These are at first of a
flesh colour, and the spines soft and hair-like. A
young one in the Shrewsbury Museum, sent by Dr.
Sankey, has the skin of a slate colour. The spines soon
become hard, and in the adult are of a yellowish-white,
with a dark band round the centre of each. They are
not straight, but bent into an elbow just before entering
the skin : this materially increases their elasticity and
prevents their being driven into the body when the
animal falls upon them from a height. Foxes and
Badgers are said to be great enemies to Hedgehogs,
and are able to destroy them in spite of their prickly
defences. Mr. Buddicom writes of a young Hedgehog,
about half-grown, which he found near Ticklerton in the
summer of 1892 :—"We brought him home and took
him into the dining-room at tea time, and fed him with
sponge-cake, which he enjoyed immensely. The little
beast showed no signs of timidity, nor did he once curl
himself into a ball. [Gilbert White says young
Hedgehogs have not this faculty.] He bit at our fingers,
drank a saucerful of milk, in which he wallowed like a
little pig, and having finished the milk, bit at the edge of
the saucer, and worried the table cloth with great
energy, just as a puppy worries clothes and things."

The Hedgehog is common all over the County, and there is a popular superstition that it sucks cows, robbing the calves of their milk. Length, nearly 12 inches. Colour, light yellowish-brown. A white Hedgehog was taken near Oswestry a few years ago.

MOLE; Provincial name, Oont. The presence of the *Talpa europæa.* Mole is shown all over the County by the abundance of molehills which are to be seen in pastures, fields, and gardens, everywhere. These heaps of earth are thrown up by the Mole in order to get rid of the soil excavated in making its burrows, but often it forces its way along just below the surface which is raised into a ridge, and no soil is then thrown out. The food of the Mole consists principally of earth worms, of which it devours immense quantities. The worms delight in moist earth and come out of their burrows at night, especially when there is a heavy dew, while in dry weather they go deep down to the moister soil below. The Mole pursues them wherever they go, and very rarely comes above ground except when the worms do. Thus it happens that very few persons ever see a *live* Mole. At times, however, the Mole will eat slugs, grubs, and other small creatures and will roam about on the surface in search of them; this happens most often on early summer evenings. For an account of a Mole pursuing and capturing a Lizard above ground see paragraph under "Sand Lizard" in this volume. It has been known to take to the water, and swim well, and will often make a tunnel to a brook or pond for it is a thirsty animal, and goes regularly at noon to the water to drink. The wondrous adaptation of the structure of the Mole to its peculiar mode of life has been remarked by so many writers that it need here only be alluded to—the immense strength of the limbs and their conversion into digging implements; the long pointed sensitive snout; the velvety fur, which will lie equally well in any direction so as to offer no resistance to motion backwards or forwards in the earth; the absence of external ears which, if present, would get clogged, and the very small eyes, which are hardly needed by an animal living in darkness. Old works on Natural History describe a central " fortress " which the Mole constructs under a bank or tree, and with which the numerous tunnels communicate, but at the present day the correctness of these statements is doubted. The males are more numerous than the females, and sometimes fight furiously. The number of young in a litter averages four or five, and the nest is a mass of leaves, roots, etc., situate beneath a large molehill. The young are blind and naked, but grow very rapidly, so that at six weeks old they are three-fourths the size of the parents and able to follow them about. Weasels will pursue and kill Moles in their burrows, and Eyton states that an old mole-catcher near Wellington caught a Mole and a *Stoat* in the same trap. Length to end of tail, nearly 8 inches. Colour, black with a grey gloss. Varieties occur of a white, cream, or buff colour; a fine specimen of the last is in the Shrewsbury Museum.

SHREW, or Shrew-mouse. Common in Shropshire, this *Sorex vulgaris.* little animal is not often seen alive on account of its seeking food only by night. In autumn large numbers of them are found lying dead in roads and lanes; there are no marks upon them, or we might suppose they got killed in fight, for

Shrews are very pugnacious. This annual mortality is
a puzzle which has baffled all attempts at solution
hitherto, though it has been enquired into by many of
our best practical zoologists. The Shrew feeds on
insects and grubs of all kinds, but does not burrow after
them. It forms little "runs" in the grass and along
hedge-banks, and in such places its shrill squeaking cry
can often be heard when the animal is invisible. In
winter the Shrew hibernates in holes in the ground,
under trees, or in other sheltered places, and its sleep
is very profound. It makes a domed nest of grass,
leaves, etc., amongst grass or in a hedge-bank, and
usually has from five to seven young. Owls eat numbers
of Shrews, and cats kill but will not eat them. Few
animals are so pugnacious as Shrews, and if two are
confined in a box they always fight to the bitter end.
All Shrews have a peculiar and characteristic odour, due
to the secretion of a pair of glands on the body, and it
was this fact possibly which gave rise to the strange
superstition alluded to in the account of the "Shrew-
Ash" given by Gilbert White in his "Natural History
of Selborne." It was supposed that if a Shrew ran
across the limb of a cow or other animal it caused agony
in that particular part, and the pain could only be
relieved by the application of the leaves of a "Shrew-
Ash." This last was an ash tree that had had a hole
bored in its trunk into which a Shrew had been thrust
alive and the hole closed with a plug! We may rejoice
that such a barbarous practice belongs to a bygone age.
In colour the Shrew varies a good deal, but is usually of
a reddish grey above, shading off to a light grey beneath.
The tail is very characteristic, being short and quad-

rangular, and covered with bristly hairs. Total length, 4¼ inches. Although Shrews are popularly classed with the Mice, they only resemble them in size and general appearance. The long tapering snout and the character of the teeth (which are of a bright red colour), show that they are not Rodents but belong to the Insectivora.

WATER SHREW. This is a larger species than the last, and *Crossopus fodiens* a very pretty little animal. As its name implies, it is found in the neighbourhood of water, in which it swims and dives readily. When swimming it appears to lie so lightly on the water that more than half the body is above the surface, and it is flattened out sideways. The fur is probably oily, for it seems to resist the water penetrating, and when the animal dives air-bubbles cling to the hairs so that it presents the appearance of a bar of silver gliding rapidly through the water. The home of the Water Shrew is a long winding burrow which it excavates in the bank of the stream or pond where it lives. The entrance is close to the edge, or even below the surface of the water, and on the least alarm the Shrew dives or swims straight to this refuge. The further extremity of the burrow is expanded into a round, grass-lined chamber, and here the female (which is smaller than the male) produces its young early in May. The litter generally numbers five or six, but sometimes more. The food of the Water Shrew consists principally of water insects and the various larvæ, snails, and crustaceans found in fresh water. It is very fond of the grubs of the caddis-fly, and will turn over stones to get at them. It has also been known to eat the spawn of frogs and fishes, and the fry of salmon, etc. It sometimes wanders away from water in search of food, but

suffers quickly from drought.   On account of its timidity
it is very rarely seen, but there is no doubt it is fairly
common in Shropshire, as specimens have been taken in
several widely-separated localities, as well as close to
Shrewsbury.   One of the best accounts of the habits of
the Water Shrew was written by a Shropshire man—the
late J. F. M. Dovaston, of West Felton—from observations made by him in 1825 on a colony he found close to
his house.   It was published in " Loudon's Magazine of
Natural History," Vol. ii., p. 219, 1829.   The fur is
fine and soft like that of the Mole, of a sooty-black
above and pure white beneath, and the transition from
black to white is abrupt—the one not shading off into
the other.   Eyton had an albino taken near Shrewsbury.
The tail and feet are fringed with stiff white bristles.
The teeth are not nearly so red as in the common Shrew.
Total length of the male, over 6 inches ; but the female
is nearly an inch shorter.   The Water Shrew is eaten by
Owls, Pike, and Weasels, and Mr. F. G. Aflalo says the
last can easily overtake it in the water !

WILD CAT.   The late Col. James Freme, of Wrentnall
Felis catus.   House, Pulverbatch, had a stuffed
                             Wild Cat, said to have been shot on
Broomhill, near that place, though the date is unknown.
There have also been several reports of the occurrence
of the animal in Shropshire, though, hitherto all
attempts have failed to elicit any indubitable evidence of
the fact, most of those reported having turned out to be
either domestic Cats that had run wild, or else Marten-
Cats (= Pine Marten), that were often known locally by the
name of Wild Cats. There is, however, little doubt that the
animal did exist in Shropshire in former days, as it was

found in other wooded districts of England, and certainly
in the adjoining parts of North Wales. It is reported
traditionally that a Wild Cat was killed after a des-
perate fight with dogs in the Corve Dale early in the
century, and there may be some truth in the report, as
the district around Ludlow is suited to its habits on
account of the extensive and secluded woods affording
safe retreat. The Wild Cat is now extinct in England,
and is only known to linger still in two portions of the
old Caledonian Forest in Scotland. Its disappearance is
not surprising, for it is a sad poacher, besides which it is
so fierce that if attacked it does not hesitate to turn on
dog or man, and so strong that it can inflict terrible
injuries with teeth and claws. It seems to be quite
untameable, for even if taken while young and brought
up in the house the kittens snarl at those who bring
them food, and pine away gradually. For this reason,
amongst others, it is now generally admitted that our
domestic race of Cats is not derived from the European
Wild Cat, though the two will interbreed. The most
obvious difference between the Tame and Wild Cat is
in the tail; in the former it is almost as long as the
body and tapering, while in the latter it is much shorter,
bushy, and does not taper at all. The Wild Cat is
also much larger, stouter built, and stronger. The
total length of a full-grown male is nearly 4 feet (of
which the tail measures only 11 inches), but the female
is smaller. The fur is always striped, and the general
colour grey ; the tail ringed with black and ending in a
black tuft, and the soles of the feet always black. The
female is lighter coloured generally. There is a stuffed
specimen in the possession of Col. Cotes, at Pitchford

Hall, and the photograph here given shows a female with three kittens belonging to Mr. Watkins, of Shotton Hall. These last were taken alive, and the keeper who shot the mother tried, unsuccessfully, to tame them. They are all Scotch specimens. The Wild Cat is easily caught in traps, as it is singularly unsuspicious, and this is the mode always adopted now for destroying it.

Wolf. We are all familiar with the fact that Wolves
Canis lupus.   formerly existed in Britain and that
they are now extinct in this country. Hume states that they were exterminated in the reign of King Edgar, who died A.D. 975. This is a mistake, for it is now known that the last Wolf was not killed in England till about the year 1500, while in Scotland it lingered till 1743, and in Ireland 1770. The following translation of a Royal Commission proves its existence in Shropshire at the end of the 13th century :—" A.D. 1281. An. 9, Edwd. I. . . . . The King to all Bailiffs, etc. Know ye that we have enjoined our dear and faithful Peter Corbet that in all the forests, parks, and other places within our Counties of Gloucester, Worcester, Hereford, Salop, and Stafford, in which Wolves may be found, that he take and destroy Wolves with his men, dogs, and devices, in all ways in which he shall deem expedient; and we command you there-fore that you be aiding and assisting the said Peter Corbet in all things that relate to the capture of Wolves in the aforesaid Counties," etc. The Peter Corbet here mentioned lived at Caus Castle. The existence of Wolves in Shropshire a century later is implied by a reference to them in " The Vision of Piers the Plowman," the author of which book was

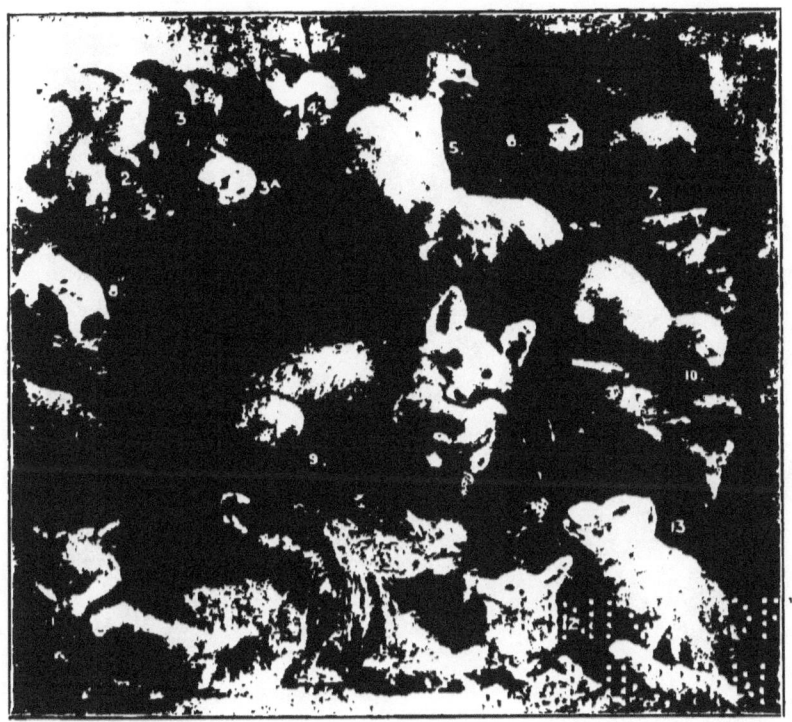

Photo by J. Franklin.   Specimens at Hawkstone.

GROUP OF CARNIVORA.

1, 2, 8. STOATS.   7. 10. POLECATS.
3, 4. WEASELS.   9, 11, 12, 13. FOXES.
5, 6. MARTENS.

WILD CAT AND KITTENS.

born in the County, near Ludlow. He says that it was
the duty of great lords to hunt in friths and forests to
find the Foxes and other beasts that are in the wild
woods and in waste places, including *Wolves*, that
worry men, women and children (see "Salopian Shreds
and Patches, 1883, p. 132 and 174). The fur of the
Wolf is long and thick, with a woolly undercoat ;
colour, reddish grey, varying between light and very
dark shades. Total length, 5 feet. It is still found
in France, Spain, Switzerland, Piedmont, and throughout Northern Europe, Asia and America, living in wild
forest land, and feeding on any animals it is able to
kill. It is of a retiring disposition, and is not known
to attack man except when much pressed by hunger,
especially in winter, when it hunts in packs. It occupies
a prominent place in literature in the familiar nursery
tale of Red Riding Hood, the ancient Fables of Æsop,
and the more modern stories of "Uncle Remus," as
well as novels, etc., dealing with hair-breadth escapes
in the frozen regions of Russia. The Wolf is really
only a species of Dog, and, though it seems incapable
of domestication, it is known to interbreed with our
domestic races of Dogs.

FOX (Fem. Vixen). In England the Fox is surrounded
Canis vulpes. by a kind of romantic halo owing to
its association with the national sport
of the country. From time immemorial it has been
adopted in literature as the type of cunning and sagacity,
and is represented in fable as invariably getting the
best of other animals by its cleverness. The wiles it
adopts to escape from the pursuing hounds have been
so often recorded that there is no need to describe them

here, and so dear are these stories to the heart of the Fox-hunter that he would be a bold man who would venture to throw discredit upon them. It may be well, however, to point out that the existence of the Fox in England is very artificial, and there is little doubt that, but for the protection afforded by the unwritten law which makes it a crime to *shoot* a Fox, the animal would be nearly or quite as rare as the Wild Cat. So much is this the case that it has influenced the whole nature of the beast, and his general behaviour is quite different here to what it is in other countries, such as Scotland, where he is shot without mercy. In Shropshire the Fox is fairly plentiful wherever there is suitable cover, and most landowners preserve it carefully. Its food consists principally of rabbits, hares, and ground birds, but it is not at all particular, and if it cannot get these, will visit the hen-roost, or eat Hedgehogs, Rats, Mice, and even Beetles. A visit to the hen-roost is disastrous, for, not content with taking one to eat, it will kill a number, and, if undisturbed, take them away and bury them for future use. The home of the Fox is known as its "earth," and is usually situated in a wood. It either excavates this itself, or takes possession of one belonging to a Badger, or enlarges an old Rabbit burrow. There is a difference of opinion as to whether the Fox and Badger will associate. Some say that where Foxes become numerous, Badgers will quit in disgust, and *vice-versa*. An intelligent and reliable keeper informed the writer, however, that he had several times found Foxes and Badgers inhabiting the same earth, and once found a Vixen with cubs in the same earth as a Badger, and, apparently, all on good terms with one another. Mr. Dumville Lees relates the following touching instance of the devotion of a Fox to its mate :— "In Scotland, where Foxes are often shot, a wounded Vixen crept away to a lonely hillside to die. It was in the winter, and the ground covered with snow. Here she was found stark dead next morning, and lying round her seven or eight animals and birds, including a Mountain Hare, while on the snow were the tracks of a Fox leading off in all directions. It was evident that the Fox had found its mate, and, unable to understand what was the matter, had spent the whole night scouring the country side in search of dainties to tempt her appetite. He must have covered a very great distance in the time, for the nearest point where he could have procured the Hare was many miles away, and at a considerable elevation." The appearance of the Fox, with his lithe red body, pointed muzzle and ears, and bushy tail, is too well known to need description. The disagreeable scent is due to the secretion of a gland under the tail. Total length, $3\frac{1}{2}$ to 4 feet. The number of cubs in a litter is from three to five. The Fox, like the Wolf, is a species of Dog, and has been known to inter-breed with the domestic species, the crosses being known as "Cocktails." Young Foxes have often been kept in kennels, but never take to a master like Dogs, and resume their wild life as soon as an opportunity to escape presents itself. A peculiarly sly look is imparted to the Fox by the oblique position of the eyes and their elliptical pupils.

PINE MARTEN. This beautiful animal is now extinct in *Mustela martes.* Shropshire, though it is still found in a few thickly-wooded parts of Wales, as, for example, near Dolgelly, where two were killed in March, 1898. Eyton, writing in 1838, says that even then it was getting rare. He mentions two taken near Stapleton. Mr. Henry Gray saw a pair at Ludlow, in 1837, that had been killed in Stokes Wood, near Onibury: These, with three others all killed in the neighbourhood about the same time, are now in the Ludlow Museum. A male and a female were killed on Wenlock Edge, about 1840, by Mr. William Pinches, of Ticklerton, and there are specimens in Harnage Grange, believed by Mr. Benson to have been killed near Lutwyche. The latest record is one killed in Bucknell Wood, in 1862, by Mr. Sitwell. The Marten (or Marten-Cat) is usually found in forest lands, and climbs trees and runs and leaps amongst the boughs with the agility of a Squirrel. In more open or hilly districts, however, it often comes to the ground, and, if hunted, retreats to the craggy tops of hills, where it takes refuge in crevices into which Dogs cannot follow it. It can travel over the ground at a rapid rate for a short distance by a succession of sidelong leaps. The food of the Marten consists chiefly of birds, which it captures by pouncing upon them unawares. It also eats Rabbits and other small Mammals, as well as Reptiles. The number of young averages four or five, and there are often two litters in a year. These are reared in the old nest of a Squirrel or Magpie. If taken young they are easily tamed, and as pets they are preferable to the other members of their tribe since they are free from any offensive odour. The general colour of the Marten is dark brown, with whitish under-parts, and a yellow patch on the throat. The tail is long and bushy, and the fur long and glossy, while there is a thick greyish under-fur. This fur is of considerable value, a good skin fetching 10s., but of course those used here are imported from the Continent. Total length, 30 inches, of which the tail forms nearly half. As already mentioned the Marten is often called, erroneously, the Wild Cat.

POLECAT. Provincial name, Fitchet; domesticated, Ferret. *Mustela putorius.* Few animals of its size are so destructive of game, poultry, etc., as the Polecat, and the result of the bad character it bears is that it is shot and trapped at every opportunity. Early in the present century it was quite common in Shropshire, and it could not be called rare in 1850, but from that time forward the number has diminished so rapidly that the time cannot now be far distant when it will be as extinct locally as the Marten. Amongst the most recent occurrences in the County are the following :— One seen in Stokes Wood, Craven Arms, in 1895; another shot at Leighton, near Cressage, in 1897; while it is reported as still existing in the neighbourhood of Bishop's Castle and about Kinnerley. Mr. T. Ruddy writes that, about 50 years ago, the Polecat was regularly hunted in the neighbourhood of Corwen. One of the animals was let out of a box on the evening prior to the hunt, and next morning—by which time it had found a retreat in some hole—the hunting party followed the cold scent with strong dogs for many a mile. In Cumberland the dogs used for the purpose were called Fell-hounds. The Polecat only comes abroad by night, lying concealed during the day in an old Rabbit hole, or in crevices amongst rocks or loose stones. Its hiding place is generally situated in a copse or on a wooded hill, though it does not often climb trees. The food of the Polecat includes almost every bird or animal it can master, with, perhaps, a preference for Rabbits. It always kills its prey by a bite on the head or neck, either penetrating the brain or opening the carotid artery, and, after eating the brains and drinking the blood, often leaves the body untouched. It is this habit which makes it so dreaded by farmers and game-preservers, for in this way it destroys far more than it can eat. It has also been known to prey upon fish and frogs, and can swim well. The colour is dark brown, with narrow whitish bands on the face. In appearance the Fitchet somewhat resembles the Marten, but has stouter limbs, and a shorter bushy tail. Length, 2 feet —the female a little smaller. The average number of young is five. The Ferret is merely a domesticated variety of the Polecat, smaller in size and lighter in colour, generally nearly white: the female is called a Jill.

STOAT, or (Winter dress) **Ermine.** This graceful, active, *M. erminea.*   and courageous little animal is plentiful in Shropshire, and more numerous than any of its tribe in spite of the numbers shot every year by gamekeepers. In its habits it resembles the Polecat, and it is quite as bloodthirsty, though not *quite* so wantonly destructive of life. A singular effect of terror is exhibited by Rabbits and Hares when a Stoat is pursuing: instead of escaping by running—as they easily might—after going a short distance they stop as if paralysed, at the same time uttering piercing squeals of distress, and await the coming of the Stoat, who kills them at once by a bite on the head. Stoats, although so cruel, are very playful, and the writer once watched a group of five on the towing-path of the Canal, near Shrewsbury, having a regular romp—chasing each other round in circles and leaping over one another with the utmost grace.   They kept this game up for a considerable time, until some movement made them aware of a man in the vicinity, when they instantly vanished through the hedge.   On examination, the body of a Stoat shows that the animal is endowed with enormous muscular strength for its size, and the length and flexibility of its body account for the extreme grace and agility which it displays in all its movements.   Besides the animals, etc., which they kill, Stoats are very fond of eggs, especially, according to Mr. de Winton, "old Stoats whose teeth are worn."   He "took 42 Pheasant eggs from one hole in May, 1894, and got the skin of the old 'Hob' who amassed this larder."   Stoats generally feed on the spot where they have taken their prey.   They can climb trees and swim readily, and the writer knows of a case where a female came down to the edge of a wide Canal uttering a peculiar squeaking call ; she plunged straight into the water, and was immediately followed by ten more, apparently young ones, though it is hard to believe they were all one family.   The most marked peculiarity of the Stoat is the change which takes place in the colour of its coat.   In cold countries the fur becomes entirely white every winter, except the end of the tail, which is always black.   In Shropshire this change is rarely complete, though all gradations are found between the pure white, pied, and brown dress.   In the white condition the Stoat is known as Ermine, and numbers of skins are imported annually from Russia and N. America to adorn the robes of Royalty, etc. The skins of British specimens are of no value, the fur being too short. There has been much controversy as to the mode by which this change of colour is effected, and it is even yet a moot point whether it is by the growth of new hair replacing the old, or by blanching of the old hair. In summer the colour of the Stoat is reddish-brown above, the under-parts yellowish-white, and the tip of the tail black. The tail is not bushy as in the two preceding species, and is of only moderate length, while the limbs are short. Length of male, 18 inches; the female smaller. The young are produced in a burrow towards the beginning of summer, and are usually five or six in number.

WEASEL. The Weasel looks exactly like a small Stoat
M. vulgaris. with a short tail, but is easily dis-tinguished by its redder colour and by the *absence* of the black tip to the tail. It also resembles the Stoat in its general mode of life, but preys upon the smaller animals. Indeed the Weasel shows a decided preference for Rats and Moles, and its small size and slender body enable it to pursue and easily capture the latter in their burrows. It also destroys large numbers of Mice and Voles, so that it is a most useful animal to the farmer in keeping these vermin in check. For this reason the Weasel should be encouraged and protected instead of being shot by keepers as it so often is; the small amount of mischief it does amongst game being far outweighed by the services rendered in the way just mentioned. Unlike most of our Mammals, the Weasel is frequently seen abroad by day, and exhibits little fear of man. The writer once pursued one in a road-side

ditch, when, after running a few yards through the grass, the Weasel ran up on the road and sat up on its haunches like a dog " begging " and steadfastly regarded its pursuer! The effect was most ludicrous. The energy and restless activity displayed by the Weasel are pro- verbial. Mr. F. Rawdon Smith writes:—" When in a strange piece of country of which it did not know " the lie " I have several times noticed the Weasel spring off the ground in order to obtain a better view, and do so repeatedly till certain of the direction it should take." There are instances on record of its having got the better of even a large Hawk that had carried it off by clinging to and biting its neck till the bird had to descend owing to loss of blood. The young are produced in a hole in a hedge-bank or similar place, and the litter usually numbers five or six. Total length of male, nearly twelve inches, but the female is much smaller and varies so greatly in size that old writers believed that there were two species, the smaller one being only half the size of the other. The tail of the Weasel is proportionately shorter than in the Stoat, while the neck is longer and more slender. In Shropshire the Weasel is common everywhere, but not so numerous as the Stoat.

BADGER. This is one of the largest of our native animals.
Meles taxus. It used to be called the " Brock," and the word is found in many Shropshire place-names, such as Brockton, Brockholes, and Brock- hurst. The Badger is of such a sly and retiring dis- position and so nocturnal in its habits that it is rarely seen in the districts where it resides, and thus it is generally regarded as a rare animal and decreasing in

numbers. So far as Shropshire is concerned, this is
certainly not the case, and it is found in most parts of the
County where there are extensive woodlands remote from
the dwellings of man. In such places it excavates its
burrow, or earth, which has only one entrance, but is
said to divide inside into two or three branches. One of
these branches leads through a hole only just large
enough to admit the body of the Badger, to a good-sized
chamber lined with dry grass which the animal rolls up
into balls, and removes, as soon as it is the least bit dirty,
replacing it with clean dry grass. The narrow entrance
to the chamber enables the Badger to defend himself
against any enemy with ease, and, knowing this, he
always rushes to it on the least alarm. The female
produces three or four young about March, and they
are blind for some time after birth. The characteristic
scent of the Badger is due to a glandular pouch beneath
the tail, but it is not so strong or disagreeable as that of
the Fox. The fact that both these animals will associate
in the same earth has already been alluded to in the
account of the Fox. The cruel sport of Badger-baiting
is now supposed to be obsolete in England. To catch
the Badger a sack was placed in the mouth of his earth
while he was "away from home" on his nocturnal
rambles. A few Dogs were then let loose in the wood
to alarm him, when he would rush to his burrow and
dash head-first into the sack, the mouth of which was
instantly closed by a running string. The poor brute
was subsequently placed in a barrel lying on its side
and a number of terriers let loose to worry it slowly
to death. The end rarely came till several of its
assailants had suffered terrible punishment, for the

Badger in defending itself would inflict frightful wounds with teeth and claws, and only eventually succumbed from exhaustion. Naturally, if unmolested, the Badger is of a most peaceable disposition, and so timid that, on the least alarm, it makes at once for its earth, for, in spite of its ungainly figure, it can run at a rapid rate. Its diet is rather a mixed one, consisting of various roots, fruits, eggs, small animals, Frogs, and insects. It is particularly fond of the grubs of Wasps, and will dig out the nests with its strong claws quite regardless of stings, against which its thick skin is proof. Opinions differ as to whether the Badger is worthy of protection or not, but there is no doubt that it does sometimes eat Pheasants and eggs, and is very fond of young Rabbits: It obtains these last by digging straight down from above, and the writer knows of an instance in which the stomach of one contained twenty-two which it had dug out of their nests. If kept in confinement the Badger is easily tamed, but most people make the mistake of giving only animal food, instead of a mixed diet as above, with the natural result —the poor beast dies. Two old and three young ones were caught alive near Kinnerley in February, 1898 (early for young ones), and kept in a farm yard. They were kept on a flesh diet, and " devoured many Fowls, but the treat they most appreciated seemed to be a Cat." The general colour of the Badger is greyish, with black under-parts, and the head white, with a black band running lengthwise past each eye. The hair is much used for making brushes, and each hair is banded red, black, and white. Total length, nearly 3 feet.

OTTER. In Shropshire the Otter is found on the Severn
Lutra vulgaris.   and most of the larger streams, and in
                                 some parts is numerous. Rev. J. B.
Meredith, of Kinnerley, writes:—" traces are constantly
seen on the Verniew and Tanat, and for two years
there has been a regular holt on the Morda brook. A
few years ago I had the pleasure of seeing one fishing
in the Tanat, and as I continued perfectly still he
seemed to have no fear, and at times came within a
dozen yards." Mr. T. C. Eyton once shot one in the
sea near Holyhead while struggling with a large Conger
Eel that had coiled round it. The food of the Otter
consists principally of fish, which it pursues in their
native element and catches with ease, being a wonderfully expert diver and swimmer. It cannot eat under
water, however, but brings its prey to the bank. If very
hungry, it will eat it on the spot, but, otherwise, carries
it off to its " holt " or lair, a hole excavated in the bank
of the stream. In eating it holds the fish down with its
fore-paws and, beginning at the shoulders, devours it
piecemeal, leaving the head and tail. It has an unfortunate habit of killing far more than it can eat, and will
often leave a fish after only one or two mouthfuls. On
land the Otter runs very rapidly, though it does not leap
or bound like the other members of the Weasel tribe.
Although its food is chiefly fish, the Otter has been
known when pressed by hunger to eat various animals,
birds, and even insects. A correspondent of Mr. Beckwith, writing from Towyn, states that Otters have
been seen to take Ducks by swimming under them and
dragging them below, and during a flood one actually
came into a farm yard after some Ducks, which it had

driven into a corner, when a woman coming out of the house caused it to retreat. The Otter occasionally gives utterance to a loud whistle, probably as a signal or call to its mate. It generally hunts by night, but in districts where it is not molested it will often remain abroad far on into the daytime. The young—three to five in number—are produced early in the summer, and remain with the parents a long time. General colour, dark brown. Length, about 3½ feet. Ears small; eyes bright; the nostrils close when under water.

QUIRREL. Unlike our other native Mammals the Squirrel *ciurus vulgaris.* comes abroad in the daytime, and for this reason it is more familiarly known than any of the others, while its beauty of form and colour, and its sprightly habits, make it a general favourite. It is found all over the County wherever there are woods, especially those that contain beech and hazel trees, on the mast and nuts of which it feeds. It also eats acorns, leaf buds, and the bark of young branches. When eating it has a pretty habit of sitting up on its haunches, with the tail elevated, and holding the food between its fore-paws. When alarmed it scampers away with marvellous agility, running along the branches even on to their slender extremities, and taking long leaps from one to another. It can also run rapidly on the ground by a succession of leaps, and if pursued always makes for the nearest tree and climbs the trunk. When it has reached a safe position on the far side of the tree it pauses and peeps round to survey the enemy. It is rare indeed to see a Squirrel miss its footing, as the feet are provided with long but strong and widely-separable toes and sharp curved claws, giving a secure grip. When dead it can be hung on to the edge of a table by its hind feet. Squirrels are very fond of eggs, and the writer knows of a case in which a number of traps baited with Thrush's eggs to catch Jays were found to capture more Squirrels than Jays. They also sometimes kill and eat small birds, but this seems to be a depraved and unnatural taste. The nest is a large structure of moss, twigs and leaves woven together with fibrous roots and grass, and is placed often in the fork of a beech tree, or on the larger boughs of a fir, close to the trunk: sometimes in a hole in a tree trunk. An interesting feature of the nest is that the entrance—on one side near the top—is elastic and self-closing. The Squirrel often builds more of these "cages" or "dreys" than it uses, and the empty ones are utilized by various birds, such as the Stock Dove, who are too lazy to build a nest for themselves. The young, three or four in number, are born in June or July. The Squirrel lays up winter stores of nuts, etc., in various holes in the ground and in hollow trees, and it does not hibernate, though in very cold weather it dozes a day or two at a time in its nest. The colour of the fur varies a good deal, but as a rule it is bright red in summer, while in winter the tips of the hairs are grey, and the red, consequently, obscured by that tint. Individuals are frequently seen with cream-coloured tails, but this is merely a passing phase, for it has been found by keeping these in confinement that at the following moult the tail reverts to the normal red hue. The tufts of hair on the ears also vary considerably in length at different seasons, and towards autumn are almost absent. Total length, including hair on the tail, 15 inches, of which the tail is nearly half.

DORMOUSE. This pretty little creature resembles the
Muscardinus avellanarius. Squirrel in general appearance and
shape, though in reality it is much
more nearly related to the Mice. It is very gentle and
retiring in its habits, and, as it rarely comes abroad by
day, it is generally thought to be rather scarce. This,
however, is not really the case, and it is to be found in
most copses by those who know how and where to look
for it. As is well known, the Dormouse sleeps through
the colder months of the year, and preparatory to this
long hibernation it makes a very thick and warm nest
of dry grasses in a hole in the ground, or in a tree or
hedge bank. In or near this it lays up a store, which
it eats when warmer weather causes it to wake, now and
then, from its winter sleep. Its food consists principally
of nuts and small fruits, though it is said also to eat
insects. It sits upon its haunches and holds the food
between its fore-paws, exactly like the Squirrel. The
young are produced in the spring, and generally number
three or four; they are at first blind, but in a few days
open their eyes and grow so rapidly that they are soon
able to shift for themselves. It is rather curious that at
first they are of a dull grey colour, and only gradually
assume the bright chesnut fur of the adult. They are
very easily tamed, and in fact, even if taken straight
from the nest, exhibit scarcely any fear of man, so that
they are most interesting pets. The eyes are prominent
and large, and the head resembles a Squirrel's, except
that the ears are round and not tufted. The body is
rather stout, and the tail of moderate length and not
bushy. The general colour is a bright chesnut above
and yellowish beneath. Total length, about 5¼ inches.

HARVEST MOUSE, This tiniest of British Mammals is
Mus minutus.   rather rare in Shropshire, and even
                        where it occurs is very local, being,
perhaps, found in one field and not in the next. Mr. T. C.
Eyton had in his collection a pair with their nest which
he found on the Weald Moors. The animal and its nest
are well described by old Gilbert White in his " Natural
History of Selborne." The nest is most wonderfully
compacted of grass woven into a perfect sphere about
the size of a cricket ball, and attached between four or
five stalks of wheat. There is no special opening for
ingress, but the grass is a little thinner in one spot,
where the Mice push their way through. Here the
female produces six or seven young, which, as in all
the Mice, are at first naked and blind, and it is believed
to have several litters in the season. Its food consists
principally of seeds of all kinds, insects and worms, and
it is said to often enter wheat stacks in great numbers,
though it more often makes burrows in the ground, in
which it lays up winter stores. " So light is the Harvest
Mouse—its weight being only the fifth of an ounce—
that it can ascend a wheat stalk and feast on the corn
in the ear; its descent being facilitated by its partially
prehensile tail. In possessing an imperfect power of
prehension in that appendage, the creature is unique
among British Mammals." [NOTE. Dr. Sankey writes:
Rats use their tails slightly in descending a stick or
twig]. The general colour is brown-red above and white
beneath, with a sharp line of demarcation. Total length
4½ inches, of which the tail is nearly half. The whole
animal is very slenderly built, and the eyes are less
full and prominent than in our other Mice.

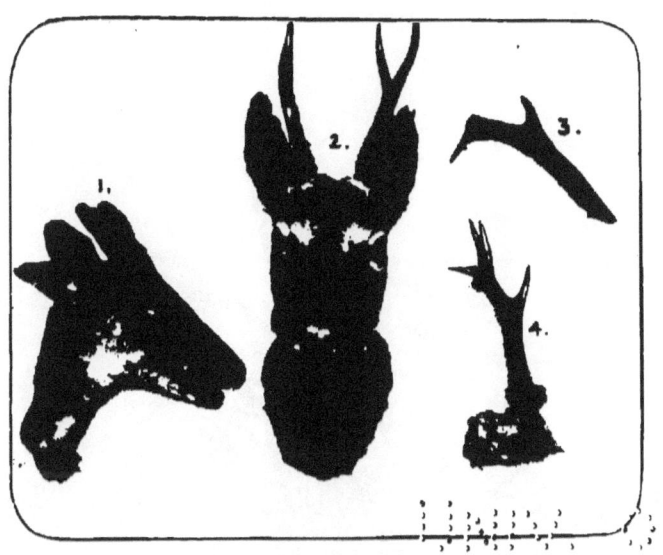

Photo by R. J. Irwin.   Specimens in Mr. G. Cooke's and Shrewsbury Museum.

ROEBUCK.

1. HEAD OF YOUNG BUCK SHOWING HORNS IN THE "VELVET."
2. HEAD OF ROEBUCK.
3. HORN SHARPENED AND USED AS A TOOL.
4. SKULL AND ANTLERS.

The two last were found in excavating foundations
for the Shrewsbury Post Office.

Photo by H. E. Forrest.　　　　　　Specimens at Radbrook.

MISTLE THRUSH AND WHITE'S THRUSH

The Harvest Mouse is reported as having been seen or taken at Bayston Hill, Pulverbatch, Broseley, and several other places which the author does not name as the records require confirmation. In the *Zoologist* for 1895 p. 447, is an interesting account of the finding of several nests at Church Stretton, and the writer, Mr. G. W. Murdock, states that this mouse is common there.

WOOD MOUSE or Long-tailed Field-mouse. This *M. sylvaticus.* little animal is abundant all over the County. In appearance it resembles the common House Mouse but has larger eyes and ears and a longer tail, while it is usually of a much lighter colour. It occurs in hedgebanks, woods, fields of grain or grass, stackyards, and sometimes even in houses. If farmers only knew the amount of newly sown seeds destroyed every year by this Mouse and the Field-vole (or short-tailed Field-mouse), they would do all in their power to lessen their numbers by protecting Owls, Kestrels, Buzzards, and Weasels, all of which prey almost exclusively on these little pests. The Barn, or White, Owl should head the list of the Friends of the British Farmer, as this bird regularly hunts over every acre of arable land for mice, and, especially when it has young, destroys them in hundreds. The Wood Mouse, although it prefers grain and seeds to any other food does not confine itself to these : it also eats bulbs, nuts, and acorns, and lays up winter stores underground. The writer has noticed a curious habit it has of biting off the " hips " or berries of the Wild Rose and storing them up in old nests of the thrush and blackbird, the crimson mass presenting a strange contrast to the bare hedge in winter. It does not hibernate, or at

E

any rate never for any length of time, so that it is preying on the farmer's pocket all the year. It generally lives in a small burrow but sometimes builds a regular nest in a hedge bank, or fills up the old nest of a bird with warm materials. It is exceedingly prolific, producing several litters in a year, each numbering from three to five young ones, and these in their turn will breed before they are six months old. The general colour is light reddish-grey above and nearly white beneath. The hind feet are very long and white. Total length about 8 inches of which the tail measures cne half. A large variety with a yellow or orange-coloured patch on the breast has been distinguished as a separate species under the name of the **YELLOW-NECKED MOUSE** (*Mus flavicollis*). It measures nearly 9 inches in length and has been taken by Mr. Dumville Lees, near Oswestry. Specimens of both are in the Shrewsbury Museum.

COMMON (House) MOUSE.   The appearance and habits
M. musculus.   of this animal are so familiar, and it
                     is so plentiful everywhere that description is superfluous. The expression " mouse-colour " is used to describe the general hue of the fur, but the range of colour in different individuals is wide. Some are so dark that they are nearly black, while others are quite a light-brown, or have grey hairs intermingled with the brown. Although it usually inhabits houses it often visits gardens and fields adjoining them, and hundreds are frequently found at the bottom of corn stacks, which are riddled with their tortuous burrows.

BLACK RAT.   There is little doubt that the true Black
M. rattus.   Rat used to occur in the County, as
                     it did in other neighbouring districts

before it disappeared on the advent of its stronger
rival the Brown Rat. At the same time careful enquiry
has failed to elicit any absolute proof of the fact.
Black Rats have been seen from time to time, and the
keeper at Betton saw one in his garden a few years
ago, but these are most likely merely black specimens
of the Common Rat. At Aldersey Hall, Cheshire, there
was a small colony of the true Black Rat, and not
many years ago one was caught and sent to the Chester
Museum : none have been seen there since. Mr. J.
R. A. Mathews reports that about 40 to 50 years
ago there were numbers of Black Rats in the buildings
on Rufford's Works, at Stourbridge. Mr. T. C.
Eyton writing in 1836 stated, that he could not hear
of any authentic case of the animal occurring in Shropshire, so that it seems unlikely that any evidence of its
existence here will now be forthcoming. It is not
known when the Black Rat was first found in England,
but it is not indigenous, and probably (like the Brown
Rat at a later date), was brought here in ships. It is
smaller and not so fierce or strong as the Brown Rat,
so that it was easily conquered by the latter. It is a
more slender animal altogether, and the tail is very long
in proportion. Length of body 7 inches ; of tail 8 or 9
inches. Colour, uniformly black.

COMMON OR BROWN RAT. As mentioned in the
M. decumanus.   foregoing paragraph, the Brown Rat
                                  was introduced here much later than
the Black species, about the middle of the 18th century.
It very quickly over-ran the Country and exterminated
the other wherever it appeared. It is now only too
common everywhere, so that a description of its habits

is unnecessary. Although it is generally found in or near to houses, the Rat often takes up its quarters in fields and gardens, or on the banks of brooks, and lives there through the summer. It swims and dives as readily as the Water Vole, and when seen in or near water the two are often confounded. As is well-known, the Rat is practically omnivorous, and when living on a stream it eats fish and snails as well as vegetable matter, while its offences are often laid to the charge of the harmless Water Vole which is a strict vegetarian. The Rat is exceedingly prolific and the numbers sometimes found together are almost incredible; over 2500 were killed in a single night in some slaughter-houses in Paris. At a farm house close to Shrewsbury, where they are not molested, they are so tame that they do not run away at the approach of man, and a policeman told the writer that he had walked up and knocked down five in one night. Compared with the Black Rat, the Brown Rat has a shorter head and more blunt muzzle, smaller ears, and shorter tail. The general colour is greyish-brown above, lighter beneath. Average length, 16 inches, of which the tail measures a little less than half. Albinos and pied varieties occur, and there is a specimen of the first, shot at Shawbury, in the Shrewsbury Museum.

FIELD VOLE, or Short-tailed Field-mouse. If the *Microtus agrestis.* Long-tailed Field-mouse is considered a pest on account of the injury it does to the farmer, what can be said of the Short-tailed Mouse? It is much more injurious, because it is usually more numerous, and at times has been known to appear in enormous numbers, causing complete devastation over entire districts. Fortunately this has never happened in Shropshire, but as a warning to those who persist in destroying such natural enemies of the mice as Weasels, Owls, Buzzards, and Kestrels, it may be well to give here details of one of the " Vole-plagues " that have occurred in Britain, prefacing it with the remark that it is only one out of many similar events in this country, while on the Continent the effects have sometimes amounted almost to a national disaster. The following is copied from Lydekker's excellent " Handbook to the British Mammalia," page 208, and refers to a district in the North of England and the Scotch border-land. " For two or three years previous to 1876, the Voles had been observed to be on the increase. In the spring of 1875 the ground, which had been covered with snow since December, was found to be riddled with holes under the wreath-drifts, and denuded of herbage, by the Voles that had found shelter there. Great numbers were seen throughout the summer, when cutting the bog hay. The Shepherd at Craikhope described the children as 'amusing themselves by hunting them from morning to night, as long as they could find nothing better to do, so that each day,' he believes, 'they destroyed hundreds, and the dogs devoured them till they made themselves sick !' In the autumn of the same year they continued plentiful. The farmer of Howpasley, when cutting a four-acre field of corn, observed numbers to be driven inwards by the reaping-machine, so that when only a spot in the centre of about twenty feet by five remained, he made one of the men take a scythe and cut it slowly, a women lifting behind. The others surrounded them, and killed the Mice as they came out; and somewhere between eighty and a hundred were thus destroyed, most of which were eaten by six dogs present. ' I used to kill scores of them,' he adds, ' with a stick while walking over the hills.' The same thing was observed, in a greater or less degree, wherever the conditions of the ground were favourable to them. A correspondent to a county paper relates that when ' removing a two-years' crop of hay in the autumn of 1875 from a meadow sloping down to the Bowmont, on the farm of Sourhope, near Yetholm, two to four nests were found under every rick, each with six to nine young ones, the nest lying in a cavity from which runs diverged in every direction. Great numbers were killed by the boys assisting. One little fellow got seventy-nine full-grown ones for his share, and his straw-hat was brimful of young ones.' Their numbers, already redundant, were augmented by the mild winter of 1875-6, and in the succeeding spring they made their presence felt in the doomed farms. During the three months from February to April they completely destroyed the pasturage of the bog-land in Borthwick water, and were then driven to the bents. Notwithstanding the means used for their destruction, which, however, were not very skilful, the swarms showed little diminution. The public journals suggested a trial of the plan which had been so efficacious in the New Forest, where holes were dug into which they fell, but the hint came too late. More efficient auxiliaries appeared in the shape of Hawks, Foxes, Weazels, &c., attracted by the abundant prey. Buzzards, which have long been strangers to the district, again made their appearance. A shepherd in Eskdalemuir saw seven of the rough-legged species

(Archibuteo lagopus) on the wing at the same time, and the short and long-eared Owls were observed in still larger numbers. By the middle of April the herbage was so much impaired that the Voles themselves began to feel the want of food, and the occurrence of severe frost, with a sprinkling of snow, about the middle of the month, completed their discomfiture. Many died of starvation, and by the end of May they had mostly dis-appeared. When the Committee of the Farmers' Club made their inspection, they found that fully one-third of the pasture in the places visited had been destroyed. The true bog-grass especially, on which the sheep mainly depend in April and May, had been eaten down to the roots. The ground was strewed with dried stalks and blades, mixed with tufts of fur, limbs, and other remains of the depredators. The sheep were in deplor-able case; several had died; and the emaciated ewes, too weak to make good nurses, suckled their lambs with difficulty. Numbers of these had perished in con-sequence, and the survivors were poor and weakly." It is scarcely necessary to add that the food of the Field Vole consists of vegetable matter of all kinds (especially grain, which it obtains by biting through the stalk above the root), and that it lives in burrows in the ground. In these it lays up winter stores of grain and berries, and feeds on them when mild weather comes from time to time, but, otherwise, hibernates through the winter uninterruptedly. The nest is in one of the burrows; the female litters five or six times in the course of a year, and has from four to eight young each time, so that it is not difficult to account for the enormous increase when their numbers are not kept in check by owls, etc. In

Shropshire the Field Vole is found in most parts that are under cultivation, and, being noticed chiefly when the grass or wheat is cut, is sometimes called the Harvest Mouse—a name which belongs properly to a very different and much rarer animal, described on a previous page. Eyton says the Field Vole frequents the banks of drains on the Weald Moors and swims and dives well. The colour of this Vole is greyish-brown above, greyish-white beneath ; length rather more than 5 inches, of which the tail measures only a fourth part. The soles of the feet are remarkable for having six pads instead of five. As in the other Voles the muzzle is short, blunt, and the whole face rounded.

BANK VOLE. In general appearance and habits this
M. glareolus. species closely resembles the last, and the two are often confounded together. It is chiefly distinguished by its much brighter colour, slightly smaller size, and relatively longer tail, while there is some difference between the molar teeth in the two species. The distribution of the Bank Vole is very partial, for, while there are large areas in which it is quite unknown, there are other districts in which it is very abundant and the common Field Vole comparatively scarce. For instance Mr. J. Steele Elliott reports that in the Wyre Forest and neighbourhood the Bank Vole is numerous, and generally more common than the Field Vole. Near Oswestry Mr. Dumville Lees has caught numbers of Field Voles but never saw a Bank Vole, while on a recent visit to Yorkshire he set a number of traps and caught none but Bank Voles. Eyton states that the Bank Vole occurred at Eyton in 1840, though two years earlier he wrote that he had never seen it.

It has not yet been found near Shrewsbury. The burrows of the Bank Vole are near the surface of the ground and it especially affects gardens near to woods ; it is fond of eating the bark and shoots of young trees, and in this way sometimes does serious damage. The nest and breeding habits are similar to those of the Field Vole.   The colour of the Bank Vole is bright chestnut above, passing to greyish-red on the sides, and whitish underneath : the feet and underside of the tail are whitish, and the tail is more thickly furred, as well as longer, than that of the Field Vole.   Total length, about 5 inches, of which the tail measures over one-third. Both species of Vole are found occasionally in gardens, even in towns, and the Field Vole has been taken in the centre of Shrewsbury.   They are certainly not desirable tenants of a garden as they root out and devour large quantities of seeds as soon as they are sown.

WATER VOLE or (as it is generally called) **Water Rat.**
M. amphibius.   Although it is numerous on all our
                                ponds, pools, canals, and streams, this
is one of the most harmless and inoffensive of animals. Yet, in ignorance of this, most people place it in the same category as Rats and Mice, and class them together as " vermin " to be persecuted and destroyed at every opportunity.   It is really a pity that the name of Water *Rat* has been so generally adopted, for the animal is not really a Rat at all, and in its habits it differs entirely from the true Rats.   The Water Vole lives in burrows excavated in the banks of a pool or stream, the entrance being usually close to the edge of the water, while there is generally a second entrance beneath the surface. If alarmed it dives instantly and travels under water

to its burrow and retires into it until it thinks the
danger has passed. Although so timid, Water Voles
will allow an observer to watch them at very close
quarters, if he keeps quite still, and the writer has many
times kept watch to see what they ate, and found that in
every case the food was of a purely vegetable character
—generally succulent stems or roots of water-plants,
such as flags, horse-tails, and pond-weed *(Potamogeton)*.
It bites off portions of these and then, sitting on its
haunches and holding them between its fore-paws,
nibbles away a little bit at a time. The Water Vole is
an expert swimmer and diver. If it is travelling along
and thinks itself unnoticed, it swims on the surface and
uses only the hind-legs to propel itself, the fore-legs
being laid back close to the body—it rarely dives except
when alarmed. Like the other Voles, the Water Vole
hibernates and lays up winter stores in its burrows.
Very rarely it is found at a distance from water, and it
has been known to live in a field, and store up potatoes,
&c., for winter use. Perhaps from the nature of its
food, the teeth are stained a yellowish-brown colour.
The fur is generally brown, but varies from quite a light
shade to dark brown. In Cambridgeshire and certain
other districts it is found almost black, and this variety
was at one time ranked as a separate species. The
Water Vole has not a great many natural enemies besides
the Weasel, but it seems that the Heron must be
reckoned as one, for, Mr. Henry Gray reports, that
not long ago, the decoy-man at Oakley Park, saw a
Heron catch one of them by pouncing on it and gripping
it firmly across the back. It then deliberately drowned
the Vole by holding it under water till it ceased struggling.

It did not eat it immediately but laid it down and watched it till, seeing signs of returning animation, it again seized it and repeated the drowning operation. The man went on with his work, and when he again looked up the Heron had bolted the Vole whole, and it could be plainly seen as a great swelling gradually descending the long thin neck of the bird! The Water Vole comes abroad by day more than most other British Mammals, but only in the morning and evening, always retiring to its burrow as long as the sun is high in the heavens. The tail is proportionately longer than in the other Voles, measuring over one-third of the total length —12 inches. The hind limbs are much longer than the fore-legs, which last are used chiefly for holding food. The female has usually five or six young which she nurses in an under-ground nest of dry grass, &c., and when they are old enough to run, they accompany her in her excursions abroad. On such occasions when threatened with danger, she has been known to convey them to a place of safety by carrying them in her mouth, just as a cat does her kittens.

HARE. The existence of the Hare in this country is, like *Lepus Europœus.* that of the Fox, to a certain extent artificial, and it is probable that, had it not been for the protection formerly afforded it by the Game Laws, it would have been ere now locally extinct, or at any rate very rare. During the few years that have elapsed since the last Ground Game Act came into force, it has almost disappeared from many areas of cultivated ground where it used to be plentiful. It still holds its own however in the more hilly parts of the County. Unlike its cousin, the Rabbit, the Hare does not live in under-ground burrows, but reposes during the day crouched together on a particular spot amongst grass, ferns, or bushes, which is called its "form." The Hare is proverbially timid, but if a man or dog approaches its form, it seeks concealment by crouching still closer to the ground, and if amongst ferns, or on a furrow, it is not at all easy of detection. If at length compelled to run it dashes off at a ;tremendous pace, progressing by leaps, as its hind-legs are much longer than the fore limbs. It can turn and double on its tracks with marvellous agility, and when pursued by greyhounds, or other dogs, always endeavours to elude them by this manœuvre, for the dogs cannot turn in so small a space, and have to describe a circle before returning on their tracks. So clever is the Hare in this way that a single dog is rarely able to catch it, and in coursing a pair of hounds are used, and when the first dog turns the Hare, the second intercepts it. When hard pressed the Hare will often take to the water and swim rapidly and well; and it has been known to swim across a river when not pursued. As remarked above, the Hare is hard to detect while it crouches on the ground, but when running the white underside of the tail makes it very conspicuous. It has been surmised that the use of this white mark is to enable the young to follow the dam in her flight, and it is noticable that if the Hare is not hurried, it travels along with the tail depressed, and then the white mark would only be visible to animals on the ground. The young of the Hare are called Leverets; generally two litters are produced in a summer, each numbering two or three. The young are born with their eyes open, covered with hair, and capable of running, and thus differ widely from those of the Rabbit. If alarmed, however, they do not run away, but behave like Partridges and other ground birds—separating and then squatting motionless on the ground. An inhabitant of open country, the Hare is found not only on the lowland meadows, but often on upland fields and moorlands at a considerable elevation, and occasionally even in woods which it enters in winter to feed on the grass which keeps green and edible in the shelter of the trees longer than it does in the exposed fields. As a rule the food of the Hare consists of grass, corn, turnips, and in fact all kinds of field crops, and it seldom comes abroad to feed till the evening. In travelling to and from its form and feeding grounds the Hare wears well-marked tracks, or paths, in the grass, and advantage is taken of its habit of always following the same route to trap it by placing wire nooses in the track. The Hare runs its head into the noose and quickly strangles itself by struggling to escape. The eyes are placed very wide apart on the sides of the head; they have large oblong pupils, and their peculiar position enables the Hare to see a dog advancing towards it from behind, while it is able to see anything on either side better than immediately in front. This is obvious if the Hare is running along a road towards a pedestrian, for it invariably comes straight towards him till within a few yards, when it will suddenly turn off through a gap in the fence on one side or the other. Its sense of hearing is very acute, and on the approach of danger it erects its long ears and turns them rapidly in all directions to detect the whereabouts of its enemy. The colour of the Hare varies considerably, but is generally reddish-grey above, and white beneath, with the tips of the ears and the top of the tail black. The coat consists of two kinds of fur, the outer one long and course, the under-one soft, short, and woolly. The fur is largely used in the manufacture of felt. The feet are covered with fur on the under-side as well as the top, the upper lip is deeply cleft, and the claws are long, curved, and sharp. Leverets are of a brighter red than the adult. Total length, about 2 feet; average weight, 8 lbs.

RABBIT. Although so similar in appearance and structure
L. cuniculas. to the Hare, the Rabbit differs widely in habits and other particulars. In the first place while the Hare is solitary, or occurs only in pairs, the Rabbit is always found in colonies—often in hundreds together. Then the Rabbit lives in burrows which it digs in the ground, and lastly its young are born naked, blind, and helpless. The colony of Rabbits with the burrows all communicating with one another are known collectively as a "Rabbit Warren." It is unnecessary to speak in detail of the habits of such a well-known animal, but, indeed, few sights are more charming to the naturalist than, seated perfectly still in the midst of a warren, to watch the playful gambols of these pretty little creatures. If unmolested they become very tame, and will not run off into their holes unless approached very closely. In Hawkstone Park there are hundreds—nay, thousands—of black and silver-grey Rabbits, and they are so used to the passing of strangers that they scarcely notice them at all, and may be seen at any hour of the day scattered like little black dots all over the green sward. The fur of these Rabbits

is of considerable value for the manufacture of black felt hats. Varieties of the Rabbit are numerous even in the wild state. Black ones have been shot at Westbury, Boreatton, and on Rudge Heath, and Eyton speaks of some very pretty buff, or yellowish ones at Longford, near Newport. At Betton, near Shrewsbury, there are large numbers of white Rabbits, and the keeper states that they have all sprung from a single white buck which he turned loose there about 18 months previously. The fecundity of the Rabbit is extraordinary, and its introduction into Australia where it had no natural enemies to keep it in check, led to disastrous results. In England this can never be the case as—besides the large numbers shot and trapped for the table—Foxes, Hawks, Weasels, and, above all, Stoats destroy very many annually. We have already, when speaking of the Stoat, described the peculiar paralysis of terror which seizes upon the Rabbit when it is pursued by that animal: it appears, however, from incidents recorded in the *Field* (May 7th, 1892, and Aug. 17th, 1897), and in the "Transactions of the Woolhope Club" (1896, p. 120), that the maternal instinct is so strong in the Rabbit that it will even overcome its fear of the Stoat. The account last mentioned states that a young Rabbit was seen by two anglers, running along, pursued closely by a Stoat, when suddenly a large Rabbit appeared upon the scene, charged furiously at the Stoat and completely bowled him over, to his great bewilderment. Upon recovering from his astonishment, the Stoat again started on the track, when the whole scene was repeated. One of the anglers, unable to contain his enthusiasm jumped up, waving his arms, and shouting, " Bravo

old'un ; Go it old'un," which had the effect of frighten-
ing the trio, and of dispersing them in different directions.
The food of the Rabbit is similar to that of the Hare,
and is entirely of a vegetable nature.   Mr. F. Rawdon
Smith says it eats bark and dandelions, though the Hare
does not.   In the severe weather of Feb. 1894 the writer
observed that the Rabbits at Bomere were driven by
hunger to climb up into the holly hedges and eat the
younger green twigs.   Although living in colonies the
Rabbit is said to be not polygamous, but to associate
in pairs, apparently for life.   [This statement hardly
agrees with the account of the white Rabbits at
Betton given above as the white buck must have
paired with several does to give rise to so numerous
a progeny in so short a time.]   The female prepares a
nest for her young in a separate burrow, lining it with fur
from her own body, and if she leaves it for any length
of time, conceals the entrance with loose earth.   The
litter comprises from five to eight young ones, which
although blind and helpless at first, grow rapidly and
are soon able to go abroad with the mother.   Several
litters are produced in a year, and the young are ready
to breed in about six months, so that they would increase
very rapidly were it not for their numerous enemies.
The Stoat and Weasel follow them into their burrows
and destroy both young and old, and the Badger will get
the young by digging down to the nest from above, while
Foxes, Owls, and other creatures attacks them in the
open.   The Rabbits is ever on the alert against the
enemy, and on perceiving one stamps loudly on the
ground with the hind feet as an alarm signal, warning
all within hearing to get into their burrows.   The skin

JACK SNIPE AND KINGFISHERS.

Photo by J. Franklin.   Specimens at Hawkstone.

GROUP OF BRITISH OWLS.

1, 3. LONG-EARED OWL.	11, 21. SNOWY OWL.
4, 5. HAWK OWL.	12, 14. TENGMALM'S OWL.
6, 7, 13. BROWN OWL.	15, 16. SHORT-EARED OWL.
2, 8, 9. SCOPS OWL.	17, 18. BARN OWL.
10, 22. EAGLE OWL.	19, 20. LITTLE OWL

and fur of the Rabbit are largely used in commerce, the former principally in making cheap imitations of more costly furs, and the latter in the manufacture of felt hats. The numerous varieties of Rabbits kept in the domestic state are all derived from the Common Wild Rabbit. The usual colour is brownish-grey above, and white beneath. Total length, about 16 inches: Average weight, 3 lbs. It is stated on good authority that this animal is not indigenous here, but the date of its introduction is quite unknown.

WILD OX. Zoologists differ in their views as to whether *Bos taurus.* all the forms of Ox found in Europe should be regarded as one or several species, and it would be out of place to discuss the question in a book of the nature of this volume. In Shropshire there are, of course, no wild Oxen now-a-days, but that they formerly existed here is proved by the fact that a bone picked up by Dr. Sankey, by the river Perry, was pronounced by Professor Alec. Macalister to be the atlas vertebra of *Bos primigenius.* Long, long ago, the natives of this country succeeded in domesticating a smaller breed of Ox to which the name of the Long-faced Ox *(Bos longifrons)* has been given. The Romans when they built Uriconium found these cattle already in the country, and probably appropriated them without scruple. That they used them for food is evident, for in the shambles found in excavating the ruins, bones of the animal are numerous. Several skulls and other parts of the skeleton are to be seen in the Shrewsbury Museum, and remains of the same kind have been found on the site of the Shrewsbury Post Office, and under the bed of the River Severn at Shrewsbury where

F

the engineers are now at work upon the new Railway
Bridge. It is believed that the original stock had
white hair, and that the celebrated Chillingham Wild
Cattle, and similar herds in other places, are lineal
descendants of this old European Ox, or Aurochs, and
that our numerous domestic races are sprung from the
same stock, but improved by a long and judicious
course of human selection.

RED DEER or Stag (male) and Hind (female). As in
Cervus elaphus.   the case of the Ox, we have to go back
                          a long distance of time to find evidence
of the existence of Wild Deer in Shropshire, and we find
it at the same place—the old Roman city of Uriconium.
Doubtless when the Romans had conquered the old inhabitants, and had had leisure to settle down and build
the substantial dwelling places, baths, forums, and shops
of which we here see the remains, they turned their
attention to hunting. The fashionable "sport" of those
days would be, not Fox-hunting, but the chase of the
noble Stag, and the active and dangerous Wild Boar.
The spoils of the chase would be borne triumphantly
home, and form the centre of attraction at the festal
board. That the Stags of Shropshire were in those
days truly a noble race, we have ample evidence in the
antlers brought from Uriconium and now to be seen in
the Shrewsbury Museum : these are all more or less
broken, but the large size of the fragments shows that
they were finer specimens than are to be met with at
the present day in the Highlands of Scotland. The
Romans used the horns for making knife-handles, and
for other purposes. Remains of the Red-deer were
also found in excavating the site of the Post Office, at

Shrewsbury (*vide* under Roedeer), and a fragment of a large antler was found 10 feet below the surface in Castle Street, in January, 1883. Most of these are smaller than the Uriconian specimens. Several of them had evidently been used as tools, and two seem to have been *sawn* off, and might, therefore, be of more modern date than the others. Antlers have also been dug up at Hawkstone, and other places in Shropshire, but there is no evidence to show at what period the animals which bore them lived. In considering the extinct animals of Shropshire it is well to bear in mind that the country was formerly in a very different state to what it is now. The County was covered with an almost unbroken and impenetrable forest and the rivers ran through extensive tracts of marshy ground. Details will be found in the introduction to this volume. This great forest afforded cover for Deer and Wild Hogs up to about the 15th century when it was nearly all cut down. The appearance of the Stag is familiar to all, even to many who have never seen one alive, through its being a favourite subject with such painters as Landseer and Ansdell. Indeed the former has perpetuated on canvas so many phases of its existence, that they almost amount to a history of the noble animal. The combats between rival males for the possession of the hinds; the gallant defence against the attacking hounds when driven to bay; the steady approach of the deer-stalker, and the wary watchfulness of the little herd on the mountain; the almost pathetic beauty of the full round eyes with large tear furrows; the timid hinds with their pretty spotted fawns; the majestic "Monarch of the Glen," proudly rearing his head aloft while his following of gentle hinds

look on as if in mute admiration—all these are depicted
with entire faithfulness to nature, yet with the additional
charm of true artistic interpretation. Little need be
added here on the natural history of the Red Deer, but
there is one point which deserves more than passing
mention—the growth and annual shedding of the antlers.
Very few facts in Natural History are so intrinsically
wonderful as this. We are so familiar with it that it
has ceased to excite surprise, but if the Deer were unknown to us, and a traveller came from a far country
bringing a pair of horns weighing fifty pounds, and
stated that the animal that bore them shed them every
spring, and grew a new pair in the course of a few weeks,
we should probably receive the statement with incredulity. The growth of the antlers is extremely rapid, and
each year another point is added and the size slightly
increased. At first only a slight protuberance or knob
is visible, covered with velvety skin : this feels hot to
the touch and there is an inflammatory action going on
beneath it which deposits bony matter continuously : the
antler keeps rapidly lengthening, still covered with the
"velvet," till it attains its due size; and then a thickened
bony ring is formed round the base of each antler, choking the arteries there and so cutting off the supply of
blood. The skin on the antlers now dries up and
shrivels, adhering to them in shreds, till rubbed off by
the stag. The antlers are shed about March, the
immediate cause of their breaking off being absorption
of bone which takes place at the point where they are
attached to the skull. The hind has no antlers. The
pairing season (at which time the stag is a dangerous
animal), is in September or October, and the fawn,

generally only one, is produced about June; the hind leaves it concealed in heather, where it lies all day, motionless, and only returns to it at night. The usual colour of the Red Deer is bright reddish-grey, and a full-grown Stag stands about 4 feet high at the shoulder, though the hinds are smaller. There are a few Red Deer in Oteley Park, but none remain in any of the other parks in Shropshire.

FALLOW DEER. This is the species which is usually
C. dama. kept in parks, but it is not indigenous here, although the date of its introduction is unknown. It is stated by some authors that it was imported by the Romans; in that case we should expect to find remains of it at Uriconium, but careful search through the specimens in the Shrewsbury Museum has failed to reveal a vestige of the Fallow Deer, though the Red and Roe Deer occur. On the other hand, amongst the remains found on the site of the new Shrewsbury Post Office (*vide* Roe Deer), are a few fragments of antlers of the Fallow Deer, associated with those of the other two species. It seems probable then that the Fallow Deer was found in Shropshire, and was hunted, many centuries ago, and possibly as far back as the Roman occupation. It will be very interesting if further excavations at Uriconium should bring to light any of its remains. On the wall of a room at New Hall, Eaton-under-Heywood, Church Stretton, is a very rough but spirited drawing (believed to have been made when the house was built in 1510), representing a hunting scene with dogs pulling down a Fallow-buck, whilst a gentleman in doublet and trunk-hose is thrusting a long spear into its neck. The following list of the parks in

Shropshire with the number of Deer in each, is copied from Mr. Whitaker's book, on the "Deer Parks of England," 1892.

Acton Burnell	180	Henley ...	70
Apley ...	150	Longnor ...	25
Attingham ...	200	Loton ...	100
Boreatton ...	70	Manor House Park	39
Chetwynd ...	115	Mawley ...	100
Hawkstone ...	300	Oteley ...	90

The Deer in these parks are all more or less tame, and are fed by the keepers in winter, but besides these we have in the neighbourhood of Ludlow, a herd of perfectly wild Fallow Deer. They reside on the Whitcliff range of Woods, the Hay Park, the High Vinnals, and the Woods of Elton and Gatley. These form an almost unbroken range between Ludford Park and Wigmore. It is believed that the Deer have had the run of these woods for centuries, but their origin is uncertain. Probably they originally sprang from animals escaped from some of the neighbouring parks. At present there are several small parties roaming the woods, and frequently three or four are seen at a time near the Wigmore road. They are, however, very shy and generally seclude themselves from observation, only decending to the lower pastures near the roads when driven by cold and hunger. Mr. Henry Gray writes:— "Some years back I was met by a very old buck that I am sure meant fight; if I had not ridden straight for him, I believe he would have come at my cob." The number of wild Deer was formerly much larger than at present, as many of them have been shot during the last thirty years. A century ago Ludford was a Deer park, and the boundary "was called the 'Deer's Leap,' because the owners, or keepers, of Ludford had the right to shoot Deer on the space outside the Park Walls, so there is no doubt the Deer often got out and in . . . . . A young fawn is occasionally found and taken by the timber-men." Mr. Gray saw one, two years ago, that was bought and reared by one of his masons. The habits of Deer are too familiarly known to need much description. Both sexes remain together throughout the year, and they are naturally gregarious. The antlers of the bucks are shed about May, and they seclude themselves as much as possible till the new ones have grown, and rub them vigorously against trees to remove the velvet. Unless taken away by the keepers, the shed antlers are eaten by the Deer, and both bucks and does share in this strange repast. The shape of the antlers is totally different from that of the last species; they are only rounded at the base, the other end being "palmate," or spread out into a broad flat expanse with points on the hinder edge only. In colour the Fallow Deer varies according to the breed and season. The old (Roman ?) breed is dark-brown in winter, while in summer it is light-red with white spots on the flanks. Another race is of a light yellowish-brown with white spots in summer, while in winter it is very dark-brown with scarcely a trace of the spots. A full-grown buck stands about 3 feet high at the shoulders; the doe is smaller. The fawns are born in June; usually there is only one, but twins are not uncommon. As venison the flesh of the Fallow Deer is held in higher esteem than that of the Stag or Roebuck.

ROE DEER. When the foundations of the new Post
Capreolus caprea. Office, at Shrewsbury, were being excavated in 1876, the workmen found
a variety of animal remains and pottery. In one spot
there had evidently been a pit dug for water, and the
sides supported by oak stakes. There is good reason to
suppose this watering place was of ancient date; that
it was made before there was any town on the spot, and
that the well remained open for many centuries. This
is a rational inference, because the pottery found in and
around the well is of all dates from Anglo-Saxon to
Mediæval times. Most of the articles found were preserved and are now in the Shrewsbury Museum. It is
clear that if the well remained *open* during many centuries,
pottery, etc., would fall into it from time to time and
so remains of all dates would become mixed up on the
bottom. This is unfortunate as, otherwise, the pottery
might have served as a clue to the date when the animal
remains associated with them, were deposited. These
last belong to the following species:—Ox, Wild Boar,
Red Deer, Fallow Deer, and Roe-buck. It is easy to
picture our hunting progenitors assembling at this well
after a hard day spent in the chase, to feast on the
spoils, building a fire of wood, roasting some joints, and
boiling others in their earthen pots. They would afterwards utilize the antlers of the Deer for various purposes.
We find one large Roe-buck's horn has been sharpened
with a knife and used as a borer, while a brow-tine off
a Red Deer has been used in the same way without being
previously fashioned. Another part of the horn of a
Fallow-deer has had a hole made through it, apparently
to nail it to a wall, and in these days we should call it

a hat-peg ! Amongst the series are several separate
antlers of the Roe-buck, and one small, but very perfect,
skull with the horns attached. In the Uriconian collection there is a handle made out of part of a Roe-buck's
horn, so that the animal was found wild here at the time
of the Roman occupation. The writer does not know
of any other remains of Roe Deer being found in Shropshire, nor has he discovered any historical evidence of
the date when it became locally extinct. We can only
surmise that it disappeared from the County about the
fourteenth or fifteenth century, for it is stated to have
lingered in Wales till the time of Queen Elizabeth. The
only districts in Britain where it now occurs wild are,
Scotland and the extreme north of England, though it has
been re-introduced into Dorsetshire, and certain other
Southern Counties. The Roe is much smaller than either
of our other Deer, and differs from them in many of its
habits. In the first place the two sexes remain together
throughout the year, and it never associates in herds; as
a rule only from two to four are found together. It is
essentially a forest-haunting animal, and may be seen
grazing, morning and evening, between the trees; it
also feeds at night, and will then sometimes descend on
to cultivated land and do some damage to grain-crops.
If alarmed it dashes off at a great pace, progressing by
a series of graceful bounds. It is of a curious disposition
and will come close up to examine anything presenting
an unusual appearance. The doe gives birth to two
fawns about May or June. The horns of the buck are
simple and straight the first winter after birth, forked
the next year, while the three tines are developed in the
third winter. After this they only gradually increase in

size with each successive season. The flesh is dark and
rather dry. Colour, reddish-brown in summer, and
yellowish-grey in winter, with white throat and a large
white patch on the rump ; the doe is lighter than the
buck, and the fawns are spotted with white along the
flanks. The male stands a little over 2 feet high at the
shoulders, but the female is smaller.

WILD BOAR. This is the progenitor of our domestic
Sus scrofa.   swine, and is indigenous to Britain, as
                           well as throughout most of Europe and
Asia. It is not known when it was first tamed, but the
hunting of the Wild Boar has been a favourite sport from
time immemorial, and is still popular in the countries
where it continues to reside. The animal is chased by a
powerful breed of dogs known as Boar-hounds, and by
men on foot and on horse-back, armed with long spears.
When brought to bay, the Boar is a dangerous adversary,
for he will suddenly face round, bristling with rage, and
charge furiously amongst his enemies, striking rapidly
right and left, and inflicting terrible wounds with his
long sharp tusks. He is generally at last overcome by
numbers, and despatched with spears. The Boar's
Head, spiced, cooked, and gaily decorated, used to be
the central dish in olden times on festal occasions,
especially at Christmas, and there is a well-known carol
that used to be sung when it was carried in

> "Caput apri defero
> Dignum laude viro," etc.

We have abundant evidence of the former existence of
the Wild Boar in Shropshire. The tusks are very hard
and durable, and numbers of them—some very large—have
been found in excavations at Uriconium, Shrewsbury,

Hawkstone, and other places, and specimens of these are in the Museum at Shrewsbury. There is documentary evidence of the existence of the Wild Boar in the neighbouring County of Stafford, as recently as the end of the sixteenth century, and it probably disappeared from Shropshire about that time. In a state of nature, this animal is less omnivorous than our tame pigs, and feeds mainly upon vegetable matter such as acorns, beech-mast, etc., but it is fond of roots, and will turn up the ground with its snout in search of them. It chiefly frequents forest-land, especially such as contain marshy hollows, going about in small droves (or "sounders,") as a rule, but old Boars are solitary in their habits. It is these old "solitaries" that afford the most exciting sport, for they are full of courage, and charge again and again without the slighest hesitation, even when severely wounded in several places. The wild sow is said to produce from six to ten young at a birth, and to have two or three litters in a year. At the present time the Wild Boar is more numerous in the Black Forest than elsewhere in Europe. The average height of an adult male is about 3 feet, and the length of the tusk (lower canine), when extracted from the jaw, about eight or nine inches. It is a curious fact that the young of the Wild Boar are marked with light longitudinal stripes, though these are rarely observed in the tame pig. The Wild Boar is also more thickly covered with bristly hair than the pig, and is of a dark iron-gray colour.

CHAPTER IV.

BIRDS.

I N possessing the power of flight, Birds are able to overcome most of the obstacles which limit the range of
strictly terrestrial animals. A large proportion of them
voluntarily undertake periodical migrations, when they
traverse hundreds, and even thousands of miles, between
the localities where they take up, respectively, their summer
and winter quarters. These various movements affect our
islands in divers ways, and make it possible to classify the
species found here in certain groups.

RESIDENTS, which do not migrate to any appreciable extent,
   such as the Sparrow, Blackbird, and Robin.

SUMMER MIGRANTS, which arrive here early in the year,
   breed here, and leave in the autumn. This group includes most of our Warblers, Swallows, the Cuckoo,
   Corncrake, and a mixed series of other species.

BIRDS OF PASSAGE which make only a temporary sojourn here
   while travelling to and from their breeding places.
   Many of the Waders come under this heading, and in
   Shropshire we might instance the Golden Plover, Spotted
   Crake, and some of the Sandpipers.

WINTER MIGRANTS. Most of the species that stay with us in
   Winter, have their breeding places in the North. The
   group includes most of the Wild Ducks, Geese, and
   Swans; Fieldfares and Redwings; Crossbills, Bramb-
   lings, and many others.

WAIFS AND STRAYS that come to us accidentally at all periods of the year. This, naturally enough, includes all the rarest birds; a few of them admitted to the British list on the strength of only two or three occurrences.

Out of the 250 species enumerated in the following pages, 87 may be classed as residents, 34 as summer migrants, 17 birds of passage, 40 winter migrants and 72 as waifs, or accidental visitors. It is difficult in many cases to say, in which category a particular species ought to be placed, and a number are here treated as waifs, which could only be so called in Shropshire, as they are resident species in other parts of Britain. Many of the sea birds are classed thus, as, for instance, the Razorbill, which is a common bird on our coasts, but has only occurred once in the County of Salop. On the other hand certain other sea birds, such as the Kittiwake, occur so frequently that they have been included here amongst the residents, and a few others that have only occurred in Shropshire in winter, are placed amongst the winter visitors. It will be understood therefore that the list is quite arbitrary, and only intended to give an approximate estimate of the numbers in each class. Shropshire has been fortunate in regard to the manner in which Birds have been observed and recorded from the early part of the century to the present time. The lists published by Eyton, Rocke, and Beckwith, leave little to be desired in the way of completeness, but the nesting and habits of our Birds, particularly the commoner species, are so full of interest that they provide an endless source of amusement and profitable study to the field-naturalist. In spite of the number of books that have been written on the subject, there are still many points in the economy of bird-life that require elucidation. For instance, it is a matter of dispute whether

certain of the Warblers and Finches are beneficial to garden
crops or not, and at the time of his death, Mr. Beckwith
was engaged in investigating the matter. Another point of
constant interest, is the observation of the arrival and departure of our summer and winter visitors, and an attempt
has been made in the following pages to show these dates
by a plan which is at once simple and easy for reference.
It may perhaps be objected that the Birds are here treated
too shortly, but that defect may be remedied in the supple-
mentary volume, which the author hopes to publish at a
later date, which will contain not only Mr. Beckwith's notes,
but a large amount of matter regarding our commoner Birds,
from the pen of the author and other local naturalists.
The account given here of the Birds of Shropshire is only
a record of all the species found in the County. It is not
intended to supersede, but to supplement any standard book,
such as Mr. Howard Saunders' *Manual of British Birds.* The
classification and nomenclature adopted are the same as in
that work. The English names of all species that reside in
Britain throughout the year, are printed in capitals. All
others are in smaller type, but the dates beneath the names
show readily which are summer, and which are winter
visitors, whilst those without dates may be treated as casual
visitors. The paragraphs printed in small type refer to
species that have occurred near to, but not within, the
County, or else to Birds recorded on doubtful evidence. In
a few instances the author has ventured to make statements
at variance with those of previous observers regarding the
relative scarcity of certain species. These cases are of great
interest as showing the changes which even a few short
years may produce in our Fauna. As an example we may
cite the great increase in the numbers of Hawfinches and

Crossbills: both of these birds used to be looked upon as rarities, whereas now the former has become so numerous in some localities, that it does serious damage in gardens; and the latter has visited us in flocks every winter lately, and a few have remained throughout the year. The Nightingale too, is slowly increasing, both in numbers and in the area of its range. On the other hand most of the Birds of Prey are getting extremely rare on account of the incessant war waged against them by man, and human agency is also responsible for the marked diminution in the number of Goldfinches. The words 'Provincial name' are here used to denote that the name is often applied in Shropshire to the species in question: they do not necessarily imply that its use is confined to this district.

The names of Resident species are printed in Capitals. Visitors and Casual wanderers in ordinary type. The average dates of arrival and departure are given below the name; the Roman Numerals indicate the *quarter* of the month. B. means that the bird has bred in Shropshire.

MISTLE THRUSH—B. Provincial name, Storm Cock; so *Turdus viscivorus*. called because it often sings loudly in stormy weather, perched on the top of a high tree. It is plentiful all the year round, but in Winter assembles in small flocks when it is sometimes mistaken for the Fieldfare. It is easily distinguished, however, from the other Thrushes by the *rounded* spots on the breast. It nests very early in the year, usually in open situations on the larger boughs of trees.

THRUSH—B. Provincial name, Throstle. Plentiful every-*T. musicus*. where and a general favourite on account of its beautiful song. It feeds largely on snails which it carries to some favourite stone in its beak, and there batters the shell to pieces against the stone, which at length bristles with the sharp fragments glued to it by the snails' slime. A beautiful variety was shot near Shrewsbury in 1898, in which the whole plumage was suffused with bright buff colour.

Redwing. A common Winter visitor, often associated
T. iliacus. in flocks with the Fieldfare. It looks
Oct. 11.—Apl. 11. like a small Thrush, but may be distinguished from it by the whitish streak over the eye and the bright red colour of the under-wings—hence its name. The Redwing can often be heard at night uttering its call note as it flies along, high overhead, and this is usually the first intimation of its arrival in autumn. It feeds almost exclusively upon insects and grubs, though Mr. Benson says that in the severe winter of 1894 several were found dead at Pulverbatch with their crops crammed full of hips and haws.

Fieldfare.—Provincial name, Feltyfare; Common Winter
T. pilaris. visitor, generally found in flocks fre-
Oct. 111.—Apl. 11. quenting fields and hedge-rows. Has been known to remain in the County as late as the beginning of May, but has never nested here. The upper part of the plumage is bluish, while the under parts are paler than in the other Thrushes, and in flight this latter feature is very conspicuous. Besides eating grubs, the Fieldfare feeds largely upon berries.

White's Thrush. A specimen of this large and very
T. varius. rare species was shot at Moreton Corbet, January 14th, 1892, and is in the collection of the late Mr. Beckwith. The photograph on page 70 shows well the boldly marked plumage and

Specimens at Clungunford.

GROUP OF FALCONS.

1, 2. GYR FALCONS.
3, 4, 7, 8, 9, 11. ICELAND FALCONS.
5, 6, 10. GREENLAND FALCONS.

Photo by R. J. Irwin.   Specimen at Mr. Irwin's.

PEREGRINE FALCON.

the large size of the bird as compared with the Mistle Thrush.

BLACKBIRD *B.* This beautiful songster is very common.
T. merula. Pied varieties occur, and a perfectly black specimen (with black legs and beak) was obtained in 1897, at Westbury, by Mr. E. R. L. Burton. Mr. Benson remembers one with a white head and neck that used to frequent the Churchyard, at Pulverbatch, about 1867-8. The female is brown instead of black, whilst the young show their affinity to the Thrushes by having the breast spotted in exactly the same way. Rev. J. B. Meredith writes:—"It is not uncommon to notice on the throat of the hen bird a ring of lighter colour suggestive of the cousinship of the Ring Ouzel·"

Ring Ouzel—*B.* Resembles a Blackbird, with a crescent
T. torquatus. shaped, white patch on the breast. It
Apl. I.—Oct. II. occurs every Summer, chiefly in the hilly parts of the County, and has been seen occasionaily in Winter. It breeds regularly on the Longmynd and hills on the Welsh border; the nest has also been found at Myddle. The nest and eggs resemble those of the Blackbird, but are always placed on or near the ground.

Wheatear—*B.* This pretty bird visits us in numbers on
Saxicola œnanthe. its Spring and Autumn migrations.
Mar. IV.—Oct. I. On arrival, the birds are so fat that if one is shot its feathers quickly become saturated with oil. Only a portion of these spend the Summer with us. They prefer low hills, and nest there in holes in the ground, frequently selecting a rabbit burrow.

G

Whinchat—*B.* Provincial name, U-tick. Common in Sum-
Pratincola rubetra.   mer wherever gorse bushes are found,
Apl. iv.—Oct. 1.   and on meadows along the Severn
Valley. Its nest is generally placed
under a furze bush. Its provincial name is derived
from its call note of U-tick, but it has also a pleasing
little song which it usually utters while perched on the
topmost spray of a gorse bush. It is easily distinguished
from the Stonechat by the broad white line over the eye.

Stonechat—*B.* Common in Summer on moorlands, but not
P. rubicola.   so plentiful as the Whinchat. A few
Apl. iii.—Oct. 1.   stay with us through the Winter in
sheltered places. It is distinguished
from the Whinchat by the uniformly black head and the
white bar on the wing.

Redstart—*B.* Provincial name, Fiery-brand-tail; so called
Ruticilla phœnicurus.   from the brightly coloured tail which
Mar. iv.—Sept. 11.   seems to "flash red" as the bird flits
by. It is a common bird in Summer
and nests in holes in trees, gateposts, walls, etc. A few
years ago a pair succeeded in rearing a brood in the
Quarry, Shrewsbury, though the nest was most conspicuously placed on a boss low down on the trunk of
a tree. These birds were watched feeding the young on
caterpillars, of which each parent brought about fifty
per hour.

Black Redstart. One killed near Wem in 1878, is in the
R. titys.   late Mr. Beckwith's collection. As its
name implies, there is a good deal of
black in the plumage, and the whole bird is darker than
the common Redstart.

ROBIN REDBREAST—*B.*   Very common.   The red-
Erithacus rubecula.   mottled eggs are almost as well known
as the bird.   Mr. G. Fox, junr., found
a nest near Shrewsbury in 1898, containing five pure
white eggs.

Nightingale—*B.*   This lovely songster does not seem often
Daulias luscinia.   to occur North of Shrewsbury, though
Apl. III.—Sept. II.   it is found sparingly every year along
the Severn Valley up to that point,
and has nested in the neighbourhood several times.
There is evidence that its range is gradually extending
Westwards, and the Rev. J. B. Meredith has heard it
three times near Kinnerley.   A nest was taken a few
years ago within two miles of Shrewsbury, but it would
not be advisable to name the locality.

Whitethroat—*B.*   Provincial names, Nettle Creeper; from
Sylvia cinerea.   its fondness for beds of nettles : Jack-
Apl. III.—Sept. I.   Straw, in allusion to its loosely built
nest of grass-stalks.   Very common.

Lesser Whitethroat—*B.*   Found all over the County, and
S. curruca.   apparently more plentiful some years
Apl. III.—Sept. III.   than others, but as a rule less numerous than the last, which it much re-
sembles, both in habits and appearance.

Blackcap—*B.*   Fairly plentiful Summer visitor, nesting in
S. atricapilla.   underwoods and shrubberies.   Its song
Apl. III.—Sept. II.   almost rivals that of the Nightingale.

Garden Warbler—*B.*   Rather common in similar situations
S. hortensis.   to the Blackcap ; it, too, is a beautiful
May II.—Sept. IV.   songster.

GOLDEN CRESTED WREN—*B.* This beautiful and
Regulus cristatus.   tiniest of British Birds is common
                              where conifers grow. The nest is
slung like a hammock beneath the bough of a Yew, or
similar tree. With the exception of the rare Golden
Oriole, this is the only British bird that builds a hanging
nest.

Fire Crested Wren. There is little doubt that this
R. ignicapillus.   bird is often confounded with the Gold
                            Crest which it closely resembles. It
may be distinguished, however, by the black line through
the eye, and yellowish lines above and below it. It has
occurred five or six times in Shropshire, no less than
three in Shrewsbury. One of these flew into a shop
on the Wyle Cop in December, 1882, and was caught.
Another shot by a boy with a catapult is in the collection
of the late Mr. Beckwith.

Chiffchaff—*B.* A common and very early Summer visitor.
Phylloscopus rufus.   It makes a beautiful domed nest on or
Mar. 11.—Oct. 1.   close to the ground, amongst rank
                          herbage, and generally ornaments the
entrance with dead leaves. The cock perches on the
topmost branch of a tall tree to utter its little song,
which consists of a reiteration of the syllables forming
its name.

Willow Wren—*B.* Provincial names, Peggy Whitethroat,
P. trochilus.   and Hay-bird; the last in allusion to
Apl. 1.—Sept. 111.   the nest, loosely constructed of dried
                            grass; it is domed at the top, and
usually placed on the ground amongst coarse herbage.
The Willow Wren is numerous everywhere. The name

"Peggy Whitethroat," is applied indiscriminately to this species, the Chiffchaff, and the Common and Lesser Whitethroats.

Wood Wren—*B.* Provincial name, Yellow Wren. The
P. sibilatrix. largest and most yellow of the three
Apl. III.—Sept. II. members of the genus. Its nest differs
from the two last in having no lining of feathers. Found in many large open woods; numerous on Pim Hill.

Reed Warbler—*B.* Occurs on most of the larger meres
Acrocephalus streperus. where the Reed (*Arundo phragmites*)
Apl. III.—Sept. III. grows, and fastens its deep cup-shaped
nest to the stems. It frequently sings at night, whence it is often mistaken by ignorant people for the Nightingale. This remark also applies to the Sedge Warbler.

Great Reed Warbler. A specimen of this rare large species
A. turdöides. was shot at Ellesmere about 1886. It
was preserved by a local man named
C. W. Lloyd, and purchased by H. Shaw; subsequently it passed through the hands of G. Cooke and G. F. Fox, to its present possessor, Mr. W. S. Brocklehurst, Kempston, Bedfordshire.

Sedge Warbler—*B.* Common. Although it prefers the
A. phragmitis. neighbourhood of water, it occurs, and
Apl. III.—Sept. IV. sometimes nests, far from any pool.

Grasshopper Warbler—*B.* A bird rarely seen owing to
Locustella nævia. its shy skulking habits, but familiarly
Apl. IV.—Sept. III. known by its peculiar note, which
resembles the chirp of a Grasshopper,

or, more nearly, the sound of a fisherman's reel.   It is found all over the County, though not numerously, and nests with us, generally preferring damp situations.

HEDGE SPARROW—*B.*   Common.   Few objects in the
Accentor modularis.   country are more familiar than this soberly clad bird, with its short sweet song, and nest with lovely blue eggs.

Alpine Accentor.   This larger and very rare species has
A. collaris.   occurred once in Shropshire, in 1891. It was caught in a brick trap at Bore-atton Park, and identified by Dr. Herbert Sankey.

DIPPER or Water Ouzel—*B.*   Common on most streams,
Cinclus aquaticus.   especially on the Welsh border.   Its beautiful dome-shaped nest, compactly built of moss, is often found on the banks, but generally escapes notice from being cunningly concealed.   An example of the Black-Bellied variety *(C. melanogaster),* was shot at Church Stretton, December 15th, 1895, by Mr. Campbell Hyslop, and reported by Mr. Paddock.

Bearded Tit.   This rare Tit seems to have been resident in
Panurus biarmicus.   former times at Aqualate Mere, on the borders of Shropshire and Staffordshire.   Sir Thos. Boughey had two eggs taken out of a nest in a gorse bush near the mere, prior to 1880, and three of the birds, from the same district, are said to have been stuffed by Harvey.

LONG-TAILED TIT—*B.*   Provincial name, Canbottlin.
Acredula caudata.   Very common.   The beautiful egg-shaped nest of moss and wool, thickly lined with feathers, and ornamented with bits of lichen, etc., is, in Shropshire at any rate, more often placed in the midst of a thicket of rose-brambles than elsewhere.   It sometimes contains

eggs by the end of March. Two broods are often reared in the year, and the whole family goes in a troupe during winter. They may then be seen roosting at night in a row along one branch of a tree.

GREAT TIT—*B*. *Parus major.* Very common. A handsome bird, noted for the habit of placing its nest in curious places, such as letter boxes, pumps, etc. The writer knows of a case where a Great Tit, having built in a letter box, was inconvenienced by the letters falling on the nest, and deliberately removed one, which was found, later on, at a considerable distance! It is called "Ox-eye," by country people, from its note, which is supposed to resemble these syllables.

COAL TIT—*B*. *P. ater.* Common in large woods; closely resembles the next species.

MARSH TIT—*B*. *P. palustris.* Not uncommon. It does not particularly frequent marshy places, as its name seems to imply. It may be distinguished from the Coal Tit by the absence of white on the nape of the neck.

BLUE TIT—*B*. *P. cæruleus.* Always called locally "Tom Tit," though the name is sometimes used for the other Tits as well. This is the commonest of all the Tits, and a general favourite, on account of its sprightly manners. Like the other Tits it is exceedingly active in its search after insects, and in pursuit of them clambers up and down the boughs in every conceivable position, as often as not head downwards, picking them out from crevices in the bark, or the recesses of leaf-buds.

NUTHATCH—*B.* Rather local in its distribution, but
Sitta cæsia. numerous in certain places (for example in the Quarry, Shrewsbury,
where it breeds every year). It is most easily seen in
winter when the trees are bare of leaves. It nests in
holes in trees, and plasters up the entrance with a very
hard clay, leaving only a small opening. It runs upwards
or downwards, or over or under branches, with equal
facility, and, as it rarely flies, except from one tree to the
next, it looks more like a mouse than a bird.

WREN—*B.* Provincial name, Jenny Wren. A pretty but
Troglodytes parvulus. shy little bird, with a wonderfully powerful and sweet song, often heard in
winter. It frequently builds more than one nest, the
extra ones being used by the cock or young to roost in.
The hen will desert on very slight provocation (if anyone
even looks at the nest), unless the eggs are near hatching
or there are young in the nest. The Wren shares with
the Robin the benefit of a local superstition, expressed
in the following lines:

"The Robin and the Wren
Are God's cock and hen."
and
"Whoever kills a Robin or Wren,
Shall never see his mother again."

On a stormy winter night, Mr. Buddicom once found the
leeward side of his house, at Ticklerton, literally covered
with wrens, apparently sheltering from the bitterly cold
wind. Mr. Ruddy relates a curious story of a Wren, at
Palè, Corwen :—She had built a nest close to the path-
way, and some workmen, passing along several times a
day, used to stop and look into the nest ; Apparently

the Wren resented the intrusion, for she closed up the opening, and made a fresh entrance to the nest on the opposite side, where she could not be overlooked!

TREE CREEPER—*B.* Not at all uncommon, but so quiet
Certhia familiaris. in its motions that it generally passes unnoticed. It is abundant in the Quarry, Shrewsbury. (Photo p. 177).

PIED WAGTAIL—*B.* Provincial name, Water Wagtail.
Motacilla lugubris. Common in Summer, and nests here regularly. The numbers are less in winter owing, doubtless, to emigration.

White Wagtail. Closely resembles the last, but has a larger patch
M. alba. of white on the sides of the neck, reaching almost to the shoulders, and a grey, instead of black, back. It has been observed several times by Mr. Paddock near Newport, but no specimen has yet been obtained for identification.

GREY WAGTAIL—*B.* More common in Winter than
M. melanope. Summer; breeds in hilly country, near small streams. Its nest has been found at Church Stretton, Oswestry, and other places.

Yellow Wagtail—*B.* A Summer migrant, pretty generally
M. Raii. distributed through the County. Often
Apl. 11.—Sept. 111. follows cattle to feed on the insects disturbed by their feet from the long meadow grass.

Tree Pipit—*B.* Common in Summer, but less plentiful
Anthus trivialis. than the Meadow Pipit. It gets its
Apl. 11.—Sept. 111. name from its habits of constantly perching on trees, though the nest is

usually on the ground amongst coarse herbage. The eggs vary in colour immensely, from dark mottled-brown to a bright rosy-red.

MEADOW PIPIT, or Titlark—*B*. Very similar in
A. pratensis.   appearance to the last, but smaller, duller in colour, and with a much longer hind claw. Common in Summer, especially on high ground. A few only winter with us. The Cuckoo often chooses the nest of the Meadow Pipit to deposit its egg in, and the eggs of both are rather alike, but that of the Cuckoo is the larger and generally has, somewhere on its surface, one or two black lines or markings.

Richard's Pipit. This rare bird is the largest of the genus,
A. Richardi.   and is distinguished by its long limbs and very long hind claw. One was killed near Shrewsbury, in October, 1866.

Rock Pipit. This is properly a shore bird, but an
A. obscurus.   immature specimen was killed at Berwick, November 23rd, 1877.

Golden Oriole. The brilliant plumage of this splendid
Oriolus galbula.   bird renders it so conspicuous that it can scarcely escape the notice of the least observant, and so it is—shot! It has occurred several times in Shropshire: Two at Harnage in 1866; One at Neen Savage in May, 1886, and several other doubtful records.

Great Grey Shrike. A rare Winter visitor, recorded eleven
Lanius excubitor.   times in the County; at Shrewsbury, Whitchurch, Harton, Acton Reynald, Hawkstone, Ludlow, Weston (Shifnal), Ellesmere, and

Westbury. A fine one was caught near Shrewsbury, in 1897, on limed twigs while in pursuit of a Chaffinch, and another shot at Monkmoor, Shrewsbury, February 18th, 1899.

Red-backed Shrike or Butcher Bird—*B.*, so called from
L. collurio.   its habit of impaling its prey—large
May I.—Aug. III.   insects and small mammals, or nest-
                                      lings—upon thorns.   It is, doubtless,
for this reason it generally selects a thick thorn-hedge or bush to nest in.   The eggs are of two types, with light-grey, or light-red spots ; there are no intermediate gradations, and the two types never occur in the same clutch.   It is found sparingly all over the County.

Waxwing.   A handsome bird, named from the curious
Ampelis garrulus.   appendages to the secondary wing
                                      feathers which look exactly as if
tipped with red sealing wax.   A rare Winter visitor, it has occured at Hawkstone, Clunguuford, Wroxeter, Ironbridge, Leebotwood, 1866, and Wellington, 1871. One was caught in a garden in Abbey Foregate, Shrews-bury, about 1858, and Mr. T. P. Blunt has a specimen shot by his brother at Underdale, about 1863.

Spotted Flycatcher—*B.* Provincial name, Miller.   Common
Muscicapa grisola.   all over the County in Summer.   It
May I.—Sept. II.   gets its name from a habit of sitting
                                      on a tree or rail, whence it darts out on
any passing insect, returning again to its perch.   The con-spicuous nest escapes detection by resembling an *old* one.

Pied Flycatcher—*B.*   Rather uncommon Summer visitor ;
M. atricapilla.   partial to hawthorn trees in parks ;
Apl. IV.—Sept. I.   breeds near Ludlow, Shrewsbury,
                                      Wroxeter, and Hawkstone.

Swallow—*B.*   Very plentiful.   The first arrivals are always
Hirundo rustica.   seen flying over water, and are few
Apl. ii.—Oct. ii.   in number, being followed by the bulk
                about ten days later.   A *White* Swal-
low was seen near Ludlow, September 17th, 1897.

House Martin—*B.*   Not so numerous as the Swallow,
Chelidon urbica.   arrives generally a few days later.   It
Apl. iii.—Oct. iii.   is this bird (not the Swallow), that
                attaches its mud nest to walls beneath
the eaves.   For some unknown reason the Martin seems
to have decreased in numbers during the last few years.
Sparrows often appropriate the nests of this bird, and
drive away the rightful owners, but a few years ago at
Buttington, near Welshpool, the Martins came in numbers, bringing plaster in their mouths, and enclosed in
the nest a pair of Sparrows that had acted in this
way, so that the nest became their tomb.   A modern
instance of an old story.

Sand Martin—*B.*   The earliest of the Swallow tribe to
Cotile riparia.   arrive.   Found all over the County
Mar. iv.—Oct. i.   wherever sand-banks occur in which
                it can burrow to make its nest.

GREENFINCH, or Green Linnet—*B.*   A pretty bird in
Ligurinus chloris.   green and yellow livery, abundant
                everywhere.   Often kept as a cage
bird, like many other Finches which follow.

HAWFINCH—*B.*   This bird is often called the Grosbeak,
Coccothraustes valgaris. on account of the size of its head and
                beak, which are so large as to be
almost grotesque.   It used to be reckoned rare, but of
late years has mulitiplied very much, nearly a dozen

nests having been found in one year close to Shrewsbury. It is very partial to peas, and if it finds a row in a garden will split the pods neatly and pick out every pea till it has finished the entire row.

GOLDFINCH—*B.* Provincial names, Seven-coloured Lin-
Carduelis elegans. net, Proud tailor, and Sheriff's man ; all in allusion to its gay colouring. Still fairly numerous in spite of bird-catchers. Its well-known partiality for the seeds of thistle and groundsel makes it favour localities where those weeds abound.

Siskin—*B.* Abundant in some Winters (as in February,
Chrysomitris spinus. 1898), when it may be seen going
Nov. III.—Feb. III. about in little flocks, while in other years very few are seen. It frequents alder trees often in company with Redpoles. A pair nested at Grinshill in 1898, but the eggs were destroyed by a Jay.

SPARROW—*B.* Far *too* numerous, owing to the ruthless
Passer domesticus. slaughter of its natural enemy, the Sparrow Hawk. Sparrows destroy quantities of seeds, both in gardens and farms, and drive away many of the birds, smaller or weaker than themselves, which are useful to man in keeping in check the many insects pests. The hen sparrow differs from the cock in having no black on the throat, or grey on the pate, whereas in the next species both sexes are alike.

TREE SPARROW—*B.* Closely resembles the preceding,
P. montanus. but, besides the distinction just mentioned, the Tree Sparrow may be known by the black patch on the *side* of the neck, which

is absent in the House Sparrow. It is frequent in stubble fields, and is not generally found near houses except in severe weather.

CHAFFINCH, or Pyefinch—*B.* Very abundant every-
Fringilla cælebs.   where.   The hen has been seen to simulate lameness to decoy a dog from its nest.   The compact nest is beautifully decorated with bits of lichen and moss.

Brambling, or Mountain Finch. Resembles the Chaf-
F. Montifringilla.   finch, but has a bright chestnut-
Nov.—Feb.   coloured patch on the shoulders.   It occurs every Winter, and some years in numbers, as in 1894-5.   A pair was seen at Ludlow in May, 1882, and at Church Stretton in July, 1898, but no nest found.   There seems to be no reason, however, why the Brambling should not breed here.

LINNET—*B.* Abundant where gorse grows; nests in the
Acanthis cannabina.   bushes.   It varies remarkably in colour and some specimens resemble the next species very closely indeed.

Mealy Redpole. Rare, or else generally overlooked.   Has
A. linaria.   been recorded at Eaton Constantine by Mr. Beckwith, at Boreatton Park by Dr. Sankey, and at Longden-on-Tern, by Rev. J. B. Meredith.

LESSER REDPOLE—*B.* Common in Winter, and a few
A. rufescens.   spend the Summer here, nesting at Shrewsbury, Hawkstone, and else-where.   It is often seen in large flocks on stubble fields during winter.

Twite, or Mountain Linnet. Very rare, except on the
A. flavirostris.   Welsh border. Mr. Lees saw a large
flock near Oswestry, April 6th, 1897.
Mr. Rocke says it nests on the Longmynd.

BULLFINCH—*B.* Not uncommon in woods, and nests in
Pyrrhula europæa.   thick bushes near Shrewsbury, etc.
Occasionally goes in large flocks,
especially in the autumn. Rev. J. B. Meredith says
the Bullfinch is very destructive to leaf-buds on Plum
trees, etc., and that it does *not*—as is often stated—
attack only those that are infested with grubs.

Crossbill—*B.* A Winter visitor of very uncertain occur-
Loxia curvirostra.   rence. Shropshire was invaded by
Dec. 11.—Mar. iv.   large numbers of Crossbills, from
December, 1894, to the end of March,
1895, and each winter since it has been numerous in the
County. A pair nested in a fir tree at Llanyblodwel, in
September, 1880, while it has been seen all through the
year in several places. A brood of quite young ones
were seen at Grinshill in 1898, and it is also reported to
have bred at Willey a few years ago. Indeed it seems
likely to become resident here if unmolested. The nest
mentioned above as occurring in September, was probably a second brood, as the Crossbill is said to be a
very early breeder, and generally has the nest ready for
eggs in February or March.

Parrot Crossbill. This large race is very rare. Mr. Rocke
L. pityopsittacus.   saw two obtained near Oswestry, and
one killed near Shifnal in 1862. Two
mentioned by Pennant as out of Shropshire, seem to
belong to this variety.

CORN BUNTING—*B.*   Often called, inappropriately,
Emberiza miliaria.   Common Bunting, for it is far from
                              common in Shropshire.   It is most
often seen in winter, searching for grain in farm yards
or fields, but occurs in summer in the neighbourhood of
Broseley, Wellington and Newport, and nests in those
districts.   It is decidedly rare in the southern half of
the county.   Rev. J. B. Meredith says that the
harsh note of this bird is of a ventriloquial nature,
seeming to come from the ground when, perhaps, the
bird is perched on a tree.

YELLOW HAMMER—*B.*   Provincial name, Writing-
E. citrinella.   master ; in allusion to the scribble-like
                              markings on the eggs.   It is also called
locally, but erroneously, the "Goldfinch."   A very
handsome bird, abundant everywhere.   Its song is said
to resemble the words " A little bit of bread and no
chee-ee-eese."

CIRL BUNTING—*B.*   Very rare, or, more probably, over-
E. cirlus.   looked, from its close resemblance to
                              the Yellow-hammer, from which it
may be distinguished by its black chin and throat.   It
is said to have nested at Ludlow, in June, 1882   In the
collection at Ticklerton is a fine male, shot at Loton Park,
and Mr. Beckwith shot one near Shrewsbury, in Jan. 1879.

REED BUNTING—*B.*   Provincial names, Reed sparrow
E. schœniclus.   and Black-headed Bunting.   Fairly
                              common by water in summer, and
about stack-yards in winter.   The nest is nearly always
on or near the ground in a swampy place, and the birds
will simulate lameness to decoy intruders from it.

HERON.

Photo by J. Franklin.   Specimens at Hawkstone.

GROUP OF WADERS.

1, 4. GLOSSY IBIS.
   2. CURLEW SANDPIPER
   3. CURLEW (YOUNG).

Lapland Bunting. Very rare, has only occurred in Britain
Calcarius Lapponicus. some forty times, generally on the
autumn migration. The collection at
Hawkstone contains two specimens killed near Shrewsbury.

Snow Bunting. A very rare Winter visitor, partial to
Plectrophanes nivalis. hilly country; has been killed on the
Oct.—Feb. Longmynd; at Habberley, Oct. 1st. 1888,
Wroxeter, Nov. 7th, 1889; and seen about Ellesmere and
Cressage. Specimens obtained here rarely show much
white in the plumage.

STARLING—*B.* Exceedingly numerous everywhere. In
Sturnus vulgaris. autumn the Starlings assemble every
evening, in enormous flocks, to roost in
the reeds by the pools at Alkmond (Shrewsbury), Croes-
mere (Ellesmere), and Tibberton (Newport). This
"gathering of the clans" is a most impressive spectacle,
and to the student of bird life is well worth a journey
to witness. Starlings are fond of following sheep and
cattle in the fields, to pick up the insects disturbed by
their feet, and will frequently alight on their backs to
pick out parasitic insects; they sometimes also catch
insects on the wing.

Rose-coloured Starling. This beautifully coloured bird has
Pastor roseus. been obtained twice in the County—
In Meole, 1841, and at Brockton,
near Bishop's Castle, 1857, both in autumn—and is reported to have been seen more recently.

Chough. One of these birds was killed near Gobowen in
Pyrrhocorax graculus. 1862. It is easily distinguished by its
red legs and curved beak.

H

JAY—*B.* Common in thick woods in spite of constant
Garrulus glandarius. persecution by game-keepers. It is
hardly to be wondered at that they
are very wary birds, and give warning of the approach
of their enemy, man, by their loud jarring cry, which
alarms every bird within hearing. They are very fond
of eggs of all kinds, and this makes them cordially hated
by keepers, who often try to kill them by laying poisoned
eggs in their way. The Jays are, however, so cunning
that they rarely fall victims (see under Squirrel page 66).

MAGPIE—*B.* This handsome bird is also common, and
Pica rustica. for a similar reason is even more
difficult to approach than the Jay.
The large nest is a wonderful domed structure of sticks,
so woven together that it is impossible to withdraw
one without breaking it. The entrance on the side is
generally protected by a fence of sharp thorns. It is
placed in a tree with little or no attempt at concealment.
Local superstition makes the Magpie a bird of omen :—

> One for sorrow,
> Two for mirth,
> Three for a wedding,
> Four for a birth.
> etc., etc.

JACKDAW—*B.* Abundant everywhere, this grey-headed,
Corvus monedula. grey-eyed bird seems to convey by his
knowing look, that he is equally at
home in town or country, venerable ruin or modern
residence. Any glittering object attracts his attention,
and many stories are told of the Jackdaw stealing
jewelry, etc., and carrying off the same to its nest or
hiding place.

RAVEN—*B.* Now very rare. It used to be seen on the
C. corax. Longmynd, Wrekin, and on the Welsh
border, and nested regularly in Hawkstone Park, on the Longmynd, and at Linley near
Bishop's Castle. The last instance was in 1884, when
some young ones were taken from a nest in a quarry,
near Church Stretton. A Raven was killed at
Ratlinghope, in December, 1895, and the Bird
has been seen near Kinnerley, in 1887 and 1899.

CARRION CROW—*B.* Still plentiful in the wilder parts of
C. corone. the County, but is very much persecuted
and is therefore gradually decreasing in
numbers. Many people speak of Rooks as "Crows,"
but the former are distinguished by their gregarious habits,
and—even on the wing—by the bare white skin round
the base of the beak. The Crow has this part covered
with black bristles. The roof of the mouth is always of a
pale flesh-colour ; that of the Rook is at first dark flesh-
colour, soon turning livid, and afterwards slate-colour.

Hooded Crow. Resembles the Carrion Crow, but has a
C. cornix. grey back and breast. Mr. Beckwith
Oct.—Mar. says it is very rare, but gives several
instances of its occurrence in the
County, three in 1889—of which two were caught in a
rabbit warren at Shirlett. It seems, however, to visit us
almost every winter but never in any numbers.

ROOK—*B.* One of the most familiar and abundant of our
C. frugilegus. birds. Space forbids details of its
many interesting traits of character,
but few birds will so well repay patient study. It exists
on insects and vegetable food, but in the severe winter

of 1894, the Rev. J. B. Meredith saw a Rook kill and eat a Thrush. Varieties are not uncommon, having white spots and patches in the plumage.

SKYLARK—*B.* Very plentiful. Its lovely trilling song
Alauda arvensis. is the delight of all who hear it and although usually to be heard while the bird is soaring is often uttered when it is squatting on the ground. This bird is killed for the table, and is very good eating, though, to the writer, it seems a thousand pities that it should ever be destroyed for this purpose. He is bound to admit, however, that the Lark does considerable damage to spring wheat.

WOOD LARK—*B.* Resembles the last, but is smaller
A. arborea. and has a much shorter tail and shorter hind claw. Its song, too, is little inferior to the Skylark's, but it more often utters it while perched, or on the ground. It occurs very sparingly throughout Shropshire.

Short-toed Lark. Very rare. The first specimen recorded
A. brachydactyla. in England was caught near Shrewsbury, October 25th, 1841. (See page 30).

Shore Lark. A bird usually found only near the sea-shore. One
Otocorys alpestris. was shot at Enville (outside the County), Dec. 7th, 1879.

Swift—*B.* Provincial name, Jack Screamer, or Squealer;
Cypselus apus. in allusion to its harsh screaming cry,
Apl. iv.—Aug. iii. uttered as it dashes on in its headlong flight. Of our Summer visitors this is almost the last to arrive, and the first to leave.

Nightjar, or Goatsucker—*B.* Provincial name, Fern Owl.
Caprimulgus Europæus. A nocturnal bird, rarely seen abroad
May II.—Sept. IV. in the daytime. Not uncommon in
ferny glades in upland woods, where it
may be seen hawking for beetles and moths, and, from
time to time uttering the curious jarring cry, from which
it takes its name. It is numerous round the Wrekin,
and near Market Drayton. The name of Goatsucker
(*Caprimulgus*), was given to it under the erroneous impression that it sucked the udders of goats and cows.

Wryneck—*B.* A shy bird with the habits of a Woodpecker
Iynx torquilla. and the colouring of a Nightjar, only
Apl. II.—Sept. III. paler. Found sparingly in summer in
the southern parts of the County, but
very rare in the northern half. It has nested several
times near Ludlow and Buildwas.

COMMON, or GREEN WOODPECKER—*B.* Provincial
Gecinus viridis. name, Yawkle; in allusion to its loud
laughing cry, usually uttered during
flight from one tree to another. A brilliantly coloured bird.
common to all those parts of the County that are well-
timbered, especially the neighbourhood of Shrewsbury,
Ironbridge, and along the Church Stretton Valley.
It does not often quit the trees, but it is said to alight
sometimes on the ground to search for ants.

GREAT SPOTTED WOODPECKER—*B.* Fairly com-
Dendrocopus major. mon. The three species usually frequent
such parts of trees as are proportion-
ate to their size. The Green Woodpecker selects the
trunk and larger limbs; the Great Spotted, the main
boughs; the Lesser Spotted, the small upper boughs.

LESSER SPOTTED WOODPECKER—*B.*

D. minor.

A small bird; frequents the topmost branches of trees, where it runs actively about in search of insects. It is often overlooked, but is not really uncommon. The following incident is related by Mr. T. Ruddy:—" A pair of the Lesser Spotted Wood-pecker paid us a visit in March, and in the following month they attempted to nest here. They made a hole nearly twelve inches in depth into the dead limb of a beech tree. One or both were at work every day for more than a week. After all their trouble they were obliged to abandon the hole to a common wren; this familiar little bird kept stuffing the hole with moss, and although the woodpecker pulled out the moss several times, the wren never gave in. Evidently the wren wished to make a comfortable and safe roosting place for itself."

KINGFISHER—*B.*

Alcedo ispida.

The most brilliantly coloured of British Birds, and therefore shot upon every opportunity! In spite of this it is fairly numerous on the Severn and its tributary brooks, and on some pools in the County. Its wonderful nest-burrow is made in the banks, and has been found even within the town of Shrewsbury. It breeds early in the year. The nest is usually lined with fish-bones, but, according to Mr. Paddock, these are only the remains of fish brought to the sitting-bird by her mate, and eaten while she is on the nest—there are no bones in the nest till incubation commences. Illustration page 87.

Bee-eater.

Merops apiaster.

A pair of Bee-eaters are said, by Decie, to have been observed near Tenbury, throughout one summer—no year is stated, but about 1875.

Hoopoe. This strikingly handsome bird, easily known by
Upupa epops.   its long crest, has occurred about a
dozen times in the County, the most
recent records being one at Market Drayton, September,
1889, and two near Ludlow, in March and April, 1895.
One of these was pursued and killed by a Hawk. A
third was shot in November the same year at Claverley,
Bridgnorth—a late date.

Cuckoo—*B.* Very abundant, though much better known
Cuculus canorus.   by its cry than by sight. Its curious
Apl. III.—Aug. III.   habits have given rise to an immense
amount of controversial literature. The
female has a curious " bubbling" note, very different from
her mate's call. A fresh egg of the Cuckoo was taken
near Shrewsbury, on 16th April, 1898, a week before the
bird had been heard. Mr. J. W. Salter says, that at
Lee Hall, near Hanwood, for three years in succession,
there have been three Wagtail's nests close to the house ;
each year there has been a Cuckoo's egg in each nest.
Mr. H. H. Hughes, of Shrewsbury, took a photograph of
one of the nests in 1898, with the young cuckoo completely filling it up. It would seem that the foster-parents
do not discover the fraud practised upon them by the
Cuckoo, even when it is repeated again and again.

BARN, or WHITE OWL—*B.* The most plentiful of the
Strix flammea.   Owls in Shropshire, and, now that
farmers and others recognise its use-
fulness in destroying mice, and do not shoot it, it is
gradually increasing in numbers. Essentially a bird of the
night, it sometimes accidentally stays abroad during the
day, when it is so bewildered by the light that it is easily

caught. The flight of this—and other species of owl—
is peculiarly noiseless. Its cry is a shrill "screech,"
hence it is sometimes called "screech-owl." (For illustrations of this, and the other British Owls, see page 88).

LONG-EARED OWL—*B*. Rather common, especially
Asio otus.   round the Wrekin. It prefers Fir woods.

Short-eared Owl. A Winter visitor frequenting moors or
A. accipitrinus.   open ground, sometimes occurring in
Oct.—Mar.   considerable numbers in such situations.
                          Rev. J. B. Meredith writes: "In the
autumn of 1874, at Tern, Wellington, we found it impossible to drive partridges, which were numerous, into
a large field of seed clover, in which we subsequently
found seven Short-eared Owls."

BROWN, TAWNY, or WOOD OWL—*B*. This is the
Syrnium aluco.   only Owl that "hoots," its cry being
                          usually expressed by the syllables
"To-whoo." It is found in thick woods, especially fir
plantations, and roosts by day crouched up against the
trunk of a fir tree, which it resembles so closely as to
be almost invisible. Owls, Hawks, and other carnivorous birds, eject the fur and bones, etc., of their food that
will not digest, by their mouths, and their nests are often
lined with these unsavoury pellets. Examination of
them proves that they feed almost entirely on rats, mice,
and large beetles. The eggs of Owls are always white
and rounded. Only two are laid at first, but before
the young hatched from these are fledged, another pair of
eggs is laid. There are instances on record of even
two broods of different ages, and a pair of eggs, being
found together in one nest.

Tengmalm's Owl.
Nyctala Tengmalmi. A specimen of this tiny and rare Owl was shot near Ruyton, in 1872, and is in the collection at Hawkstone.

Eagle Owl.
Bubo ignavus. A splendid bird which occasionally wanders to Britain from the forests of Northern Europe. One was caught at Steventon, Ludlow, about 1868, and kept alive for three years: it is now in the Ludlow Museum. Another was killed near Bridgnorth in 1873; and a third near Onslow in 1887—now in the possession of Mr. Barrett, Cross Gates Inn. All three may be birds escaped from aviaries. The photograph on page 88, fig. 10, shows an Eagle Owl in the extraordinary attitude it assumes when irritated or enraged.

Marsh Harrier.
Circus æruginosus. The birds that follow are classed together under the name of BIRDS OF PREY. They are distinguished by their sharp curved talons, hooked beaks, and a membrane called the "cere" over the nostrils and top mandible. The females are, in all cases, larger than the males. The Marsh Harrier was never numerous, and is now very rare in Shropshire. Specimens have been obtained many years ago, at Berwick, and on the Longmynd, and one was seen by Mr. Dumville Lees, near Oswestry, in January, 1886.

Hen Harrier—B.
C. cyaneus. Almost as rare as the last, but has been shot at Ticklerton (a pair about 1840), Clungunford, Wem, Whitchurch, Ludlow (1887), and Ruabon (1892). A male was seen by Mr. Dumville Lees, near Oswestry, in 1879, and in 1894 a female was seen for several days about Betton Pool.

The male and female are so widely different that they were formerly regarded as distinct species. The prevailing colour of the former is slate-grey; of the latter, mottled-brown. A nest with four eggs was taken on Shawbury Heath, May 25th, 1890, and the eggs and hen bird are now in the possession of Mr. Jas. W. Lloyd, of Kington, Herefordshire.

Montagu's Harrier. A female was shot at Petton about
C. cineraceus. 1860. All the Harriers seek their prey while skimming along just above the ground, which they "quarter" like a trained spaniel.

COMMON BUZZARD—B. A handsome bird, and as it
Buteo lagopus. is one of the largest of our birds of prey, it rarely escapes the gun of the gamekeeper. It was formerly common, and even now a year seldom passes without one or more being sent in to the bird-stuffers from different parts of the County. The Buzzard can be distinguished on the wing by its habit of flying round in spiral curves. It is rather lazy, however, and often remains perched on a tree for a long time. It is a great pity that the Buzzard is being so rapidly exterminated, as it feeds chiefly on Voles, and is very slightly injurious to game—see page 75.

Rough-legged Buzzard. A rare wanderer, visiting us from
B. lagopus. Northern Europe in Autumn or Winter. It gets its name from having the legs feathered down to the toes. Used to occur near Ludlow, Pontesbury, and the Stiperstones, but the most recent records are at Wytheford (1871), a pair near Ellesmere (1877), Moston, near Hawkstone (1889), at Stretton (1895), and at Weston (1895)—see page 76.

White-tailed Eagle. This fine bird—the only eagle that
Haliaëtus albicilla. has occured in Shropshire, for the
Golden Eagle has never wandered
hither—has occurred at least eight times in the County,
most, if not all, being immature. The most recent were
at Hawkstone, March 1883, Bucknell 1892, and Craven
Arms, November 7th, 1896.

SPARROW HAWK—*B.* In spite of unrelenting persecution
Accipiter nisus. this beautiful Hawk is certainly the
most numerous of its tribe in Shrop-
shire, and breeds here regularly. It is partial to the
neighbourhood of woods, and preys chiefly on small
birds, which it seizes as it skims rapidly along the hedge-
rows, but alights (usually on the ground) to eat.

Kite—*B.* Old writers speak of the Kite as of quite common
Milvus ictinus. occurrence, but it must now be regarded as very rare. Mr. Beckwith
wrote in 1879, that a few still tried to nest near Ludlow,
and that the bird had also been seen on the Breidden.
One was shot at Wallop, October 25th, 1887, and
another at Bucknell, November 4th, 1895. Mrs. Rocke
says Kites used to build yearly in Stokes Wood, near
Craven Arms, and two or three nests of live birds have
been taken there.

Honey Buzzard—*B.* A rare visitor, recorded about ten
Pernis apivorus. times—between 1865 and 1872. One
was shot in the Edge Wood, Sep-
tember 19th, 1875. Mr. Ashdown received a pair which
had been shot in Herefordshire while in the act of
devouring a wasp's nest. Their stomachs were full of
wasp grubs, and they must previously have rifled a bee's

nest, for their beaks were encrusted with wax. In a
M.S. book of Mr. Rocke's is a note to the effect that
a pair of Honey Buzzards were taken in Ferney Hall
Dingles, June 2nd, 1865; both were trapped at a
pheasant's nest within a few hours of one another.
When first seen they were in the act of destroying the
eggs, five of which they had carried out and broken.
On dissecting the female is was found that the ovaries
were much enlarged, one egg was ready for extrusion,
and there was little doubt that one or two eggs had
been laid previously. Mr. Shaw also stated that it
had bred in Shropshire. (*Ibis*, 1865, page 13).

PEREGRINE FALCON—*B.* This handsome bird was
Falco peregrinus.   the one chiefly used in the ancient
                                 sport of Falconry. It breeds still in
North Wales, especially along the coasts, and every year
a few appear in Shropshire, generally immature birds, in
the autumn, and more females than males. No recent
instance is known of its nesting here. The Peregrine
Falcon shown on page 106, was shot near Shrewsbury,
while in the act of devouring the partridge on the
ground at her feet. One of those in the Hawkstone collection was shot while perching on the top of St. Chad's
Church, Shrewsbury, eating a Coot which it had killed.

Iceland Falcon. Mr. Rocke states that a pair of these fine
F. islandicus.   Falcons was obtained near Leebot-
                              wood, about 1860. The beautiful group of birds photo-
graphed on page 105 is at Clungunford, and shows
specimens of the Iceland, Greenland, and Gyr Falcons,
in various states of plumage—a case of birds that is
probably unrivalled.

Hobby—*B.* Formerly often obtained near Shrewsbury, but
F. subbuteo. now rather rare. It only visits this
Apl.—Oct. Country in summer and leaves in
autumn. It has bred in Shropshire.
It never builds a nest for itself, but uses that of a magpie
or other large bird. The Hobby has soft plumage and
very long wings, which, when folded, almost exceed in
length the tip of the tail.

Orange-legged Hobby, or Red-footed Falcon. Mr.
F. vespertinus. Rocke mentions an immature bird in
Mr. Bodenham's collection, caught near
Shrewsbury, about 1868, and there is an adult in the
collection of Mr. Chase, (Birmingham), shot by a lad
while scaring birds near Ellesmere, in 1873.

MERLIN—*B.* A small but bold little Falcon, still occurring
F. æsalon. not unfrequently on the Longmynd
range, and along the Welsh border,
especially in winter. It rarely breeds in Shropshire,
but a nest was found on the Longmynd, in 1896. The
Merlin is particularly fond of larks, and about 1840
a pair was captured, near Shrewsbury, in a lark-net.

KESTREL—*B.* Often called "Windhover" from its habit
F. tinnunculus. of hovering stationary in the air while
scanning the ground in search of mice,
etc., a difficult feat of "wingmanship," attained to by
few other birds. The Kestrel is resident; fairly
numerous where it is not molested, and breeds regularly
in Shropshire, generally utilising the old nest of some
other large bird. It is partially migratory, as we have
fewer with us in winter than in summer. The cry is
like the mew of a kitten. It feeds principally on mice,

and rarely kills partridges, or other birds. The writer is strongly of opinion that the Kestrel, and its eggs, ought to be protected by law.

Osprey. The only one of our birds of prey that feeds
Pandion haliaëtus.   almost exclusively on fish, which it captures in its claws, plunging upon them from a great height. In Shropshire, it has been obtained at Chetwynd in 1833, Clun 1841, Petton 1858, Colemere 1863, Caynton 1865, Shifnal, October 1881, Willey, April, 1888, and on the Isle Pool, October 30th, 1889. Its appearance is very striking on account of the large amount of white in the plumage. The beautiful photograph, of which the frontispiece of this book is a reproduction, was taken by Mr. John Franklin. The birds, which are in his possesion, were preserved and mounted by his father, Mr. W. Franklin.

CORMORANT. The two species of Cormorant and the
Phalacrocorax carbo.   Gannet are essentially sea birds that wander now and then as far inland as Shropshire—nearly always young birds. The Cormorant has been obtained at Clungunford, Atcham, and other places. One was shot near Shrewsbury in September, 1897. Strange to say, a colony of fourteen Cormorants used to live on a small island called "the Bylet," in the Severn, at Fitz, near Montford Bridge. They were usually to be seen perched on the top of some tall ash trees, from whence, now one and then another, might be seen to dash downwards into the water. They were there from 1820 to 1839, but the number dwindled, one by one, till there were only seven, when some Rooks came and took possession of the trees ; the Cormorants then left and

never returned. Old residents in the neighbourhood can still remember these birds, but, though they were constantly under observation they were never known to nest.

SHAG, or Green Cormorant. Has been reported less
P. graculus. often than the above, but immature birds have occurred at Ellesmere, Hawkstone, Longville, and—in September, 1897,—at Polmere.

GANNET, or Solan Goose. A large bird which when
Sula Bassana. adult is chiefly white, but the young is brown with white flecks. When in pursuit of food, which consists of fish near the surface of the sea, they fly along low down to obtain velocity, when shooting aloft till the momentum is exhausted, they plunge with closed wings into the sea, a great shower of spray flashing upwards around the spot. As they fish thus in large flocks, the spectacle is most exhilarating. The Gannet has occurred at Market Drayton, near Shrewsbury, and near Clee Hill. One was found exhausted at Ruyton-xi-Towns, in 1890, and three at Stapleton, April 24th, 1896, one of which was killed.

HERON—*B.* Frequently seen on rivers and pools, standing
Ardea cinerea. motionless in the shallows till some unlucky fish swims by, when it is darted upon with unerring aim. (For an account of the Heron eating a Water Vole, see page 80). Some of the old Heronries are now deserted, but the Heron still breeds at Attingham, Oakley, and Walcot Parks ; Colemere, and Halston. At the last named place they have been seen, in a gale, holding on to the boughs by their beaks. Illustration page 123.

Squacco Heron.   A specimen of this little Heron was
A. ralloides.   killed near the Brown Clee, in 1834, and
                        is now in the collection at Ticklerton.

Night Heron.   An immature bird was shot many years ago
Nycticorax griseus.   near Wroxeter, by Mr. Stanier.

Little Bittern.   One mentioned by Pennant was shot in the
Ardetta minuta.   Quarry, Shrewsbury, early in the cen-
                       tury.   One occurred at Crosemere, 19th
May, 1880; and one at Ditton Priors, 29th May, 1894, now
in Mr. Watkins' collection.   In July, 1881, a pair was sent
to Shaw for preservation ; shot at Petton and Marbury.
The female was full of eggs, and would doubtless have
bred here if unmolested.

Bittern—B.   This handsome bird is reported by Eyton to
Botaurus stellaris.   have bred at Cosford, near Shifnal, in
                        1836.   It is now very rare, but single
birds still occur from time to time.   In 1895 a female
shot near Oswestry, November 28th, and a male at
Marbury, December 21st ; as these places are not far
apart these were, possibly, a pair.

Glossy Ibis.   This handsome purple-black bird has occurred
Plegadis falcinellus.   only once in Shropshire—a pair near
                          Sundorne in 1854.   They would have
been unnoticed but for the following curious circum-
stance.   A countryman while waiting his turn to be
served, in the shop of Mr. W. Franklin, in Mardol,
was looking at the cases of stuffed birds, and
seeing amongst these a Curlew, remarked that he had
shot one the previous day, only his was *black !*   Mr.
Franklin guessing what the bird must be, offered a
liberal sum if the man would bring it to him ; the man

Photo by Jones & Son, Ludlow.   Specimens at Clungunford.

GROUP OF WILD GEESE.

———

1. RED-BREASTED GOOSE.
2. BRENT GOOSE.
3. EGYPTIAN GOOSE.
4. BERNICLE GOOSE.
5. BEAN GOOSE.

6. WHITE-FRONTED GOOSE.
7. CANADA GOOSE.
8. PINK-FOOTED GOOSE.
9. GREY-LAG GOOSE.

GROUP OF DUCKS.

1, 2, 4. LONG-TAILED DUCK (Y., F. and M.)
5, 9. FERRUGINOUS DUCK (F. and M.)
6. SURF SCOTER.
7, 8. SCAUP DUCK (F. and M.)
10, 11. BUFFEL-HEADED DUCK (F. and M.)
12. COMMON SCOTER.
13, 14. TUFTED DUCK (M. and F.)
15, 16. GOLDEN EYE (M. and F.)
17, 18. VELVET SCOTER.
19, 20, 24. HARLEQUIN DUCK (2 M. and F.)
21, 22, 23. RED-CRESTED POCHARD (Y., M and F.)
25, 26. POCHARD.

said he expected the pigs had eaten it, but that there were two of them, and he would try and shoot the other. Next day he brought in the Ibis, the pigs had *not* eaten it, but the second bird had escaped. (Illustration page 124).

Grey Lag-Goose. This is the species from which our
Anser cinereus. domestic race of geese is supposed to
Nov.—Mar. have sprung. It is very rare in Shropshire, and no specimens have been obtained recently. Mr. Henry Gray reports that in the winter of 1855 a solitary Grey Lag was seen about Bromfield for three weeks, whilst on December 24th, 1890, a flock of thirty-two visited the same locality, and remained in the neighbourhood, in small parties, for about two months. This bird is shown in company with the other species of British Geese on page 141. The plate is taken from a photograph of the handsome case at Clungunford.

White-fronted Goose. A winter visitor of uncertain
A. albifrons. occurrence: Two were obtained near
Oct.—Mar. Ludlow, 1855, one on the Teme, December 14th, 1871, and two killed out of a flock of eight at Ruyton-xi-Towns, January 1st, 1891.

Bean Goose. More numerous than the other wild geese.
A. segetum. Three were shot out of a flock of eight
Oct.—Mar. at Oakley Park, in 1861; two out of twenty at Wroxeter, in 1878; one out of a large flock at Coalport, in 1881; one at Kinnerley, February 18th, 1888; and one obtained at Lutwyche, January 17th, 1891.

I

Pink-footed Goose. A smaller bird than the last, and
A. brachyrhyncus.   occurs less frequently in Shropshire.
Oct.—Mar.   One was obtained on the Tern in 1842,
        two seen (one shot), at Eyton, 1879,
and one shot near Oswestry, February 18th, 1894. It
is worthy of note that in all geese the sexes are alike in
plumage, while in most ducks they differ greatly.

Egyptian Goose. This bird is often kept in a semi-
Chenalopex Œgyptiacus. domesticated state on ornamental
        waters, so that there is some doubt
whether it may be admitted to the list as a wild species.
One was obtained near Shrewsbury, and another at
Hatton, Shifnal, in the winter of 1878-9. In 1884 one
was shot, out of three, on the pool at Halston, Oswestry,
and in the same year one was observed by Mr. H. Gray,
near Ludlow.

Canada Goose. From time to time specimens of this fine
Bernicla Canadensis.   bird are seen in an apparently wild
        state, as, for instance, on Bomere Pool,
in February, 1896, but it is so often kept in confinement
that these are almost certainly escaped birds. Mr. T.
Ruddy reports that large numbers of apparently wild
birds have resided of late years on a farm near Runcorn.
They are shy of strangers, but will come at the call of
the farmer, who feeds them. They go away often for
several days at a time, but always return. It is more
than likely that the Canada Geese seen in Shropshire
are individuals that have wandered from the head-
quarters at Runcorn, while out on one of these excursions, for they do not stay at any one place for longer
than a day or two. Besides these casual wanderers,

however, a certain number of these Geese constantly resort to Hawkstone, Combermere and Ellesmere, and there are generally seven or eight nests on the mere at the place last named.

Bernacle Goose. Very rare, but has occurred in the
B. leucopsis. County many years ago. It is this
Oct.—Mar. bird that was fabled to be bred from the Bernacles or 'Barnacles' found attached to ships timbers.

Brent Goose. Although this bird visits the eastern counties
B. brenta. in large numbers, it is rare in Shrop-
Oct.—Mar. shire. One was shot near Shrewsbury in 1861, another on Combermere (over the edge of the County), November 5th, 1895, and one near Kinnerley, in October, 1898.

Wild Swan, or Whooper. Now very rare, but in 1837
Cygnus musicus. Mr. H. Shaw received no less than
Oct.—Mar. twenty-five for preservation. A young male was shot near Linley, January 16th, 1891. Mr. Brownlow Tower writes that several Whoopers were seen on Ellesmere Mere and the pool at Halston, in the winter of 1892-3; two were shot, of which one is now at Halston. The cry of this swan is a loud "whoop" uttered on the wing—hence its second name. The tame Swan is silent.

Bewick's Swan. Like the last now very rare. Mr. Rocke
C. Bewicki. had one, killed on the Severn about 1862.

TAME, or MUTE SWAN—*B.* Common on many orna-
C. olor. mental waters in a domesticated condition, and numerous on the meres

at Ellesmere, and at Attingham, in a half-wild state. A pair used to nest under the English Bridge, in Shrewsbury, but never reared the nestlings; the eggs were generally destroyed by floods or rats, and at length the birds went away altogether.

SHELDRAKE. This very handsome duck is a shore bird, *Tadorna cornuta.* but is often kept in a tame state on ornamental pools. Specimens occur from time to time in the County; possibly escaped from confinement, but in the winter of 1884-5, a flock of eleven wild birds visited the Severn, and several were shot at Cressage, Buildwas, and Melverley. The Shelduck and drake are shown on page 142, figs. 18 and 19, and all the other species of British Ducks on the same page, or on page 159.

WILD DUCK, or Mallard—*B*. This is the species from *Anas boschas.* which our domestic breeds of Duck originated. Many nest with us regularly, and more numerously during recent years, but in the winter we are visited by numbers from the North. This bird nests very early in the year amongst sedges, etc., by the water side, and has been known to lay eggs at Christmas. It often inter-breeds with the tame and other species of duck, and it is a curious fact that many other wild species of duck inter-breed, the resultant hybrids being very puzzling to the ornithologist. The Wild Duck has been known to nest occasionally in trees, utilizing the old nest of some other bird.

Gadwall. Has occurred only twice in Shropshire—a specimen in Mr. Bodenham's collection, *A. strepera.* killed on the Severn; and another at

Winsley, shot by Mr. J. Whitaker, November 1st, 1889, and now in his possession. This latter bird when first seen had a mate.

Shoveler Duck—*B.* Easily recognised by its curious wide *Spatula clypeata.* and flat beak, whence it is called the "Spoonbill" (a name only properly applied to one of the Heron tribe *Platalea leucorodia*, that has never occurred in Shropshire). The inside of the beak resembles a comb, and "the object of so singular an apparatus is to sift water and mud for the sake of securing the insects and worms they contain." The Shoveler is mostly an uncertain Winter visitor to this County. Mr. Rocke observed a small party on a pool at Hopton Heath during the winters of 1876 and 1877, and one was killed at Uffington about that time. Mr. H. F. Harries reports that the Shoveler Duck used to nest on Shrawardine Pool. Mr. Paddock says that a few remained at Weston (Shifnal), all through the summers of 1895-6-7. A very young bird was shot by Mr. Salter, at Hencote, in August, 1869—evidently reared there.

Pintail Duck—*B.* Has occurred several times in winter, *Dafila acuta.* near Ludlow, and other parts of the Oct.—Mar. County. One was shot at Monkmoor, in January, 1892, and a young male near Shrewsbury, in October, 1898. The bird takes its name—and is easily distinguished by—the two long narrow feathers in the centre of the tail. A pair nested in Oakley Park, in 1881.

TEAL—*B.* This is the smallest of our Ducks and seems to *Querquedula crecca.* be getting more numerous. It breeds regularly on many of our pools and bogs, making its nest of rushes, etc., and lining it with down from its breast. The cream-coloured eggs, lying on their bed of dark-grey down, are very pretty. The Teal is remarkable amongst ducks for displaying devoted affection for its young. Besides the resident birds, large numbers come to us every winter from the north. The flight of the Teal is very rapid.

Garganey—*B.* A very rare duck occurring chiefly on spring
Q. circia. and autumn migration ; none have been seen in Shropshire lately. A nest was found near Shrewsbury, about 1888, and identified by H. Shaw, from the eggs and lining of down, as belonging to the Garganey.

Wigeon. A small duck, not uncommon during winter,
Mareca penelope. though of late years it has become
Sept.—Apl. less numerous than formerly. It nests rarely in Cheshire, near our boundary, but not in Shropshire, so far as is known. Mr. Ruddy reports that in 1898 it nested at Palé, Corwen.

Pochard, or Red-headed Poker—*B.* This and the
Fuligula ferina. following ducks are expert divers,
Oct.—Mar. whereas the foregoing swim on the surface and only submerge their heads in search of food. The Pochard frequents most of our larger pools during winter, often in numbers, and occasionally nests in Shropshire, as at Tong, in 1875.

Tufted Duck—*B.* Another fairly common visitor, dis-
F. cristata. tinguished by its black crest and light-
Oct.—Mar. blue beak. It has nested in Shropshire in several places. Four broods were reared at Sandford in 1891, and it has bred every year

since 1880, at Weston Park, Shifnal. Mr. Brownlow Tower well describes these pretty ducks on Ellesmere Mere, as busy birds, "almost lifting their bodies into the air in their attempts to make a good header, and obtain some luscious insect, or tempting portion of vegetable matter."

Scaup Duck.
F. marila.
Nov.—Mar.

Named after its peculiar cry. It is decidedly rare in Shropshire. The two sexes differ widely in plumage. Young birds have been obtained on the Severn, Alkmond Pool, and at Newport. An old male was shot on Shrawardine Pool in 1895, and a nearly adult male at Kinnerley, January 6th, 1899, out of a flock of about fifteen.

Golden Eye.
Clangula glaucion.
Oct. 11.—May 1.

Not uncommon during severe winters, especially on the Severn. The male is a handsome bird with green head (slightly crested), black back, and white under-parts. Mr. Brownlow Tower writes that the Golden-eye " rises from the water with great rapidity, his wings producing a noise from which he is often known as Rattle-wing, and Whistler."

Long-tailed Duck.
Harelda glacialis.
Nov.—Apl.

A rare visitor. One was killed on Tong pool, November 6th, 1871. The ' long tail' from which the bird takes its name is seen only in the male.

Black Scoter.
Œdemia nigra.
Oct.—Mar.

As its name implies, this duck is black all over. Its visits to Shropshire are very irregular, but it has occurred at Burrington, Ellesmere, Cound, Buildwas, and Caynton. Five, out of a flock of twenty, were shot on the pool in Acton Burnell Park, November 24th, 1892, and a male near Iron-bridge, in November, 1896.

Velvet Scoter. Resembles the last but has a white patch
Œ. fusca. by the eye and on the wing. It has
Oct.—Mar. only been obtained once in the County, near Shrewsbury; the specimen was in Mr. Bodenham's collection.

Goosander. This and the two following species belong to
Mergus merganser. a group called the Saw-bills, having
Nov.—Mar. long narrow beaks serrated along the edges, and terminating in a hook. They are very expert divers, and swim fast under water. When on the surface they sit low in the water, carrying the head with the beak immersed and fishing as they go. The Goosander is a large handsome bird, and occurs almost every winter on the Severn and the larger pools. In 1876 a flock of sixteen were on the pool in Hawkstone Park for two months—two were shot; and in the winter of 1885-6 there were several flocks of ten, fifteen, or twenty, on the Severn, between Atcham and Cressage— several were shot.

Red-breasted Merganser. Resembles the preceding, but
M. serrator. is rather smaller, and has a wide red
Nov.—Mar. band on the breast. Much less common than the Goosander. Several were shot near Clungunford many years ago, a fine adult male at Minsterley, in November, 1889 (now at Mr. Palin's, Meole), and a female at Burway, near Ludlow, February 23rd, 1895.

Smew.
M. albellus.
Nov.—Feb.
This smallest of the Saw-bills is a very irregular visitor to Shropshire. Most specimens obtained here are young birds, but Mr. Rocke had an adult pair killed near Shrewsbury, and a . male at Clungunford. A specimen in the collection at Hawkstone was killed in the park. An adult female was shot at Melverley, January 22nd, 1892, and three females were noticed by Mr. Henry Gray, on the Onny, at Bromfield, in January and February, 1891. Mr. Brownlow Tower saw a male and two females on Ellesmere mere a few winters ago.

WOOD PIGEON—*B.*
Columba palumbus.
Local name, Quice or Quist; also called Ring Dove, from the band of white feathers on the neck. Exceedingly numerous in all our woods, and has increased during the last few years. It is very destructive to grain, peas, and seeds. During the severe weather of January-February, 1895, the Wood-pigeons were driven by hunger to resort to gardens, and were to be seen devouring cabbages and sprouts with avidity close to the windows of houses. Its pleasant "cooing" voice is familiar to all. (Illustration page 160, fig. 3).

STOCK DOVE—*B.*
C. œnas.
Very common, though not so numerous as the last. In flight this bird is distinguished from the Wood-pigeon by the general slate-blue colour, and the absence of the white collar. Its breeding habits, too, differ; for it makes no nest, but lays two white eggs in a hole in a tree or cliff, or in the old nest of a squirrel, owl, or other bird. Its note is shorter and more like 'grunting' than 'cooing.' (Illustration page 160, fig. 4).

ROCK DOVE—*B.*   This bird is the ancestor of our
C. livia.   domestic races of pigeon, and is often
                                     called the Blue Rock Pigeon, from its
colour, and from the habit of nesting in crevices of
rocks.   It resembles the Stock Dove in most particulars,
but in flight is easily distinguished by the white axillaries
and under wing-coverts.   (In the Stock Dove these are
grey).   The rump also is white instead of grey.   Mr.
R. B. Benson found a colony of these birds, about
1884, nesting in some deep narrow caves, in " Ipikin's
Rock," a detached portion of Wenlock Edge.   He took
two of the eggs; hatched them and reared the young
under tame pigeons to be sure of the species.   Subsequently some Jackdaws took possession of the caves
and ejected the pigeons, though specimens have been
shot in the neighbourhood since 1886.   Mr. Dumville
Lees reports that a Rock Dove was shot near
Oswestry, early in 1898; and that it used to breed
in some rocks at Treflach.   Major A. Heber-Percy
has a large number of the true breed in a domestic
state, at Hodnet.   It is difficult to say with certainty
whether or not those found in the open are escaped
tame birds reverting to the wild state.   (Illustration page
160, figs. 5 & 6).

Turtle Dove—*B.*   Provincial name, Wrekin Dove.   The
Turtur communis.   smallest of our pigeons, and, unlike
Apl. iv.—Sept. iv.   the other two, only with us in Summer.
                           It is fairly common in the County,
especially round the Wrekin, and increasing annually.
In habits and nesting it resembles the Quice, but in
flight is easily distinguished by its brown colour and
small size.   (Illustration page 160, fig. 1 & 2).

Sand Grouse. This singular Asiatic bird, has, on several *Syrrhaptes paradoxus.* occasions, visited Britain in numbers, and a few even nested in England. The only records for Shropshire are: two killed out of a large flock near Oswestry, and a flock of twenty seen near Ludlow, in 1863; one killed against telegraph wires near Craven Arms, and several others shot in the neighbourhood of the Clee Hills, in 1888.

BLACK GROUSE—*B.* (Blackcock, male, and Greyhen, *Tetrao tetrix.* female). Numerous in the southern, but rare in the northern parts of the County. The lyre-shaped arrangement of the tail-feathers in the cock is very remarkable. It breeds in the parts where it occurs, and hybrids between the Black-cock and Pheasant have occurred several times. Black Game seem to have been much more numerous here fifty years ago than they are at present, and, though it is hard to give any reason for it, they are gradually decreasing in numbers. Black Game prefer wooded hills, while Red Grouse only flourish on heathery moors.

RED GROUSE—*B.* Remarkable amongst Birds as being *Lagopus Scoticus.* the only species which is confined to the British Isles, though there are closely allied species on the continent. The result of careful enquiries in the County proves almost conclusively that there were few, if any, Red Grouse in Shropshire before 1840. About that year Mr. William Pinches and Mr. Buddicom imported two pairs from Yorkshire, and turned them out on the hills above Church Stretton (see page 27). The birds increased gradually, though it was several years before any number were shot. At the present day, however, bags of forty and fifty brace are not uncommon, and even ninety brace have been recorded. From the Longmynd, as a centre, the Red Grouse have spread to other hills in Shropshire, and they are now found firmly established on the Clee Hills, and Clun Forest, and stray birds are met with on the Ercall Hill, and in other strange places. Of course it cannot be stated with certainty that these have all sprung from the Yorkshire birds, as there is no barrier to prevent the influx of others from the Radnorshire and Berwyn mountains, where there have always been Red Grouse. Furthermore it is difficult to imagine why the Longmynd should have been devoid of this species before 1840, seeing how eminently suitable the district is to its habits. Grouse are subject to a peculiar disease, the nature and cause of which are but little understood.

PHEASANT—*B.* Common everywhere, owing to the protection of the game laws. The true old *Phasianus Colchicus.* English breed is characterized by the dark colour of the whole neck. It is gradually disappearing by interbreeding with the next species, and pure-bred *P. colchicus* are getting scarce. The Illustration on page 177 shows a male and female of the true breed, and the dark throat and neck of the former is very evident, while the light buff and brown plumage of the hen is equally conspicuous.

RING-NECKED PHEASANT.—*B.* This species was introduced into England about a *P. torquatus.* century ago. It is distinguished from the preceding by its generally lighter colouring and by a band of white feathers in front of the neck. In Shropshire it now far outnumbers the old breed, but pure-bred birds of either species are rare.

PARTRIDGE—*B.* Very common, and too well-known to need any description. White specimens occur sometimes, and in August, 1887, there were five seen in a covey of twelve at Ensdon.

Perdix cinerea.

RED-LEGGED PARTRIDGE—*B.* A much more brightly coloured bird, rare in Shropshire. In 1877 one was killed at Charlton Hill and another at Middletown. It has also been obtained at Churchstoke, near the Breidden, Weston Park, and at Willey, but at the last two places the eggs were introduced and hatched by Common Partridges. It was thought that they would be good sporting birds but they turned out quite the reverse. When driven they run rapidly along the ground instead of rising; and they fight with and kill the Common Partridge.

Caccabis rufa.

Quail—*B.* Resembles a diminutive partridge. In Shropshire it may be regarded as a regular summer visitor, though never numerous. Five were shot at Munslow, Sept. 5th, 1893. It probably nests often in the county, though the only recorded instances are: at High Ercall in 1878—a nest with thirteen eggs; at Waters Upton in 1881, (reared the young); at Montford Bridge in 1884,—a nest with seven eggs; in 1885 a nestling found near Shrewsbury and sent alive to Mr. Beckwith; and a nest with eighteen eggs found in 1893 at Hinnington, Shifnal, and identified by Mr. Paddock.

Coturnix communis.
Apl.—Oct.

Virginian Colin. A specimen of this American Quail is in the Hawkstone collection. It was shot on the estate

C. Marylanda.

but how it got there is not known. It is not a British species, and must have escaped from confinement.

Corncrake, or Landrail—*B.* Plentiful, especially in
Crex pratensis. lowland meadows. It is always re-
Apl. iv.—Sept. iv. garded as a summer visitor, and its
well-known "crek-crek" is almost as generally recognized as the harbinger of summer as the advent of the first swallow. It is probable, however, that a few birds remain here through the winter, as specimens have been shot at Leighton on the 21st Dec., 1894; at Harley on Oct. 5th, 1895; at Bolas in Dec., 1896; and at Hawkstone in the month of January. Although so often heard, the Corncrake is seldom seen, as it is very shy and lives habitually amongst long grass and coarse herbage. When injured it has been known to simulate death.

Spotted Crake. Easily distinguished from the last by its
Porzana maruetta. smaller size and light-spotted plumage.
Apl.—Oct. It occurs in Shropshire from time to
time on its spring and autumn migrations, but has never nested here. The most recent specimens are: one killed by flying against telegraph wires at Rednal, April 9th, 1892, and one shot near Minsterley Nov. 17th, 1898.

WATER RAIL—*B.* A very shy bird, frequenting quiet
Rallus aquaticus. reaches of the Severn and many pools,
and concealing itself on the least alarm.
For this reason it is rarely seen except in winter, when there is little cover to protect it. It is not really uncommon, and two nests have been found in the County—one by Mr. H. F. Harries, near Yorton, in 1881, and the other by Mr. G. Fox, Junr., near Shrewsbury, May 2nd, 1896.

MOORHEN, or WATER HEN—*B.* Very abundant on
Gallinula chloropus. our ponds, and, unlike the Water Rail,
not at all shy, for, if unmolested it
becomes very tame and will frequent lawns, etc., close to
houses. About 1876-7 there were some very pretty pied
Waterhens in Attingham Park. Rev. J. B. Meredith
says that they eat eggs, for a large number died from
eating poisoned eggs placed to destroy Magpies and
Crows.

COOT—*B.* Local name, Bald-headed Coot, in allusion to
Fulica atra. the white forehead which contrasts
strongly with the black plumage. The
Coot is not quite so abundant as the Water Hen, and
while the latter is found on small ponds, the former con-
fines itself chiefly to large pools. In hard frosts the Coot
resorts to the Severn. A curious light-brown variety was
shot at Newport in 1894.

Crane. Mr. Rocke mentions a specimen of this rare and splendid bird,
Grus communis. shot at Trippleton on the Teme, by Mr.
Roberts of that place. It was cooked and eaten !

Great Bustard. The occurrence of this fine bird in Shropshire is
Otis tarda. somewhat doubtful, but is recorded here on the
evidence contained in a letter written by the
late Rev. R. W. Gleadowe of Frodesley, to Mr. Beckwith, in May,
1879. It stated that his father-in-law, Mr. T. L. Gleadowe, saw a
pair of Great Bustards on the Longmynd. " He was riding from
Church Stretton to Ratlinghope, about July in 1826, and saw two
birds, apparently a pair, among the fern and ling about half way
between the two places mentioned. They only remained a moment
or two and then took to flight. His description is that they were
like a turkey hen in shape, of a chestnut colour above and ashy grey
beneath, but the neck was more of the same thickness throughout than
a turkey's. He used to be very fond of natural history and, to judge

by other instances, his observation and memory are careful and accurate, so that I see no reason, except the rarity of the birds, to doubt his account." The illustration on page 178 is taken from the specimen referred to on page 28.

Little Bustard.   Another rare bird that has occurred once
O. tetrax.   only in Shropshire, at Edgmond, in the
                              spring of 1883. This specimen was
sent by Mr. Paddock to Mr. Beckwith, and is still at Radbrook.

Stone Curlew, or Thick-knee. A nocturnal bird not
Œdicnemus scolopax.   uncommon in Southern and Eastern
Apl.—Oct.   England, but only once recorded (by Mr.
                            Rocke) in Shropshire, on Ponsart Hill.

Dotterel. Fairly common in some of the hilly parts of
Eudromias morinellus.   England ; this little Plover is very rare
Apl. iv.—Sept. i.   in the County. In the collection at
                          Ticklerton is one killed by Mr. William
Pinches on the Longmynd about 1840. Mr. R. B. Benson shot one at Lutwyche in 1871, while on May 12th, 1886, three were killed out of a flock of thirteen near Wellington, and identified by Rev. W. Houghton. The remaining ten mounted high into the air and flew away at a great rate.

RINGED PLOVER. Found principally on the shores
Ægialites hiaticula.   wherever they are flat, this pretty little
                           bird with its conspicuous black band
on the chest, is only a casual visitor to Shropshire, known to have occurred four or five times on the Severn in the neighbourhood of Cressage and Ironbridge. Both eggs and young resemble the shingle on which they are found.

Photo by Jones & Son, Ludlow.   Specimens at Clungunford.

GROUP OF DUCKS.

1, 9.   TEAL.
2, 3, 4.   SHOVELER (2 M. and 1 F.)
5, 6, 8, 13.   GARGANEY (1 F. and 3 M.)
7, 10.   GADWALL (F. and M.)
11, 12.   PINTAIL (F. and M.)
14.   AMERICAN WIGEON.
15, 16.   WIGEON (F. and M.)
17.   RUDDY SHIELD-DRAKE.
18, 19.   COMMON SHIELD-DRAKE.
20, 21.   WILD DUCK AND SIX DUCKLINGS.

Specimens at Hawkstone.

GROUP OF PIGEONS.

1, 2. TURTLE DOVES.   4. STOCK DOVE.
3. WOOD-PIGEON.   5, 6. ROCK DOVES.

Golden Plover. A pretty bird with yellow and black
Charadrius pluvialis. mottled plumage, occurring almost every
Oct.—Mar. winter in Shropshire on its autumn and
spring migrations, sometimes numer-
ously—as in the spring of 1879, and winter of 1895-6.
A flight of about 150 was seen on the Weald Moors in
Oct. 1889. It nests on moorlands, but is not known to
have bred in the County. In the breeding season the
breast and under parts are black.

PEEWIT, or Lapwing—*B.* Often called simply "Plover."
Vanellus vulgaris. The first name is from its well-known
cry, the second from its slowly-flapping
flight. "Plovers' eggs" are in great demand as a delicacy
for the table. Always four in number and arranged with
their pointed ends together, they are laid in a slight hollow
in the ground. If anyone approaches the place the
parent birds betray great anxiety and strive to lure the
intruder away. As in all birds of this order, the young
are able to run as soon as hatched, and, if alarmed,
separate, and seek safety by squatting close to the ground.
Peewits pair off in March, but during the winter go in
large flocks, occasionally numbering a thousand or more.
They are found chiefly on lowland meadows, or other open
country, and are most useful in destroying noxious grubs.
They suffer severely in hard frosts.

Turnstone. Essentially a shore bird, it is not surprising
Strepsilas interpres. that the Turnstone is very rare in
Aug.—May. Shropshire. One was killed near
Atcham many years ago, and a young
bird, shot at Rednal, about 1851, is now in the possession
of Mrs. Gell, of Bayston Hill.

J

OYSTER-CATCHER. Another shore bird, often called
Hæmatopus ostralegus.   Sea-pie, on account of its pied plumage.
           It is plentiful on the neighbouring
coasts of North Wales, and has occurred in the County
at Atcham, Wem, and Cruckton.   Mr. Brownlow
Tower observed a pair at Ellesmere, on the 29th March,
1899, and watched their movements through a field-glass
for some time.   The long straight bill is bright yellow,
while the legs are pink and the feet have no hind toe.
(Illustration on page 195).

Avocet. A long-legged, black and white bird, with curious
Recurvirostra avocetta. awl-shaped beak.   Pennant in his
           British Zoology, (1812), says that the
Avocet sometimes occurred on the Shropshire meres.   It
is now very rare—even in the Fen district.

Grey Phalarope. Distinguished from the Dotterel and
Phalaropus fulicarius. other small plovers, by its slightly
           webbed toes, this pretty grey and white
bird visits us frequently on its migrations, generally in
autumn.   It is very confiding, and seems to resort to
ponds, often close to farm houses.   It is equally at home
on land or water.   The most recent occurrences are: one
shot on Ludlow Racecourse Oct. 7th, and another killed
against telegraph wires, near Walcot, Oct. 5th, both in
1896.   Shropshire specimens are almost invariably in
the light winter plumage.

Red-necked Phalarope. A much more rare bird than the last.
P. hyperboreus.   Mr. Paddock in a letter to Mr. Beckwith states
           that a pair was shot near Newport in the
autumn of 1890, but omits the species from his book, so the record is
here indicated as doubtful.

Woodcock—*B.* This well-known bird comes to us in
Scolopax rusticula. "flights" in October, but a few pairs
Oct.—Mar. remain here to nest in the neighbourhood of the Wrekin, Baschurch, and
Ludlow. The Woodcock has a habit of traversing certain
tracks through the woods, when flying forth to feed in the
evening, and returning to its covert in the morning.
There are several "Cockshuts" (= cock shoot) in Shropshire—supposed to be so called because the sportsmen
waited in these places to shoot the "cocks" as they flew
by. Mr. W. E. Edwards while walking some years ago
in a thick wood behind the Whit-cliff at Ludlow, surprised
a Woodcock with a brood of several young ones. She
picked one of them up and flew away, carrying it held by
her beak pressed to her breast. When flying the Woodcock does not hold up its head but carries it low, with the
beak pointing downwards.

SNIPE—*B.* Provincial, Full Snipe. Another bird dear to
Gallinago cœlestis. sportsmen, and difficult to hit on
account of its peculiar zig-zag flight.
The extraordinary drumming noise which it makes while
on the wing, during the breeding season, has given rise to
much controversy and is alluded to at some length in
Darwin's "Descent of Man." The Snipe is found on the
Longmynd, and in most boggy places, and a few nest with
us in marshy situations.

Great Snipe. Provincial, Double Snipe. An autumn visitor
G. major. which has occurred several times in
Shropshire. Mr. Paddock shot a pair
near Newport, during the hard frost of 1878-9. This species
is distinguished by the large amount of white in its tail.

Jack Snipe. Never so plentiful as the common Snipe, the
G. gallinula.   Jack Snipe is an irregular winter visitor
                            to Shropshire. Occasionally it is seen
also in early summer but has never nested with us.
(Illustration, page 87.)

Bonaparte's Sandpiper. The first specimen of this rare
Tringa fuscicollis.   species obtained in Britain was shot at
                            Stoke Heath, before 1839. It is now in
the collection at Hawkstone.

Dunlin. A common shore bird that varies remarkably in the
T. alpina.   colour of its plumage at different
                            seasons. It rarely visits Shropshire,
but has been seen on the Teme, and one was shot at
Ticklerton, about 1840; another at Hawkstone, about 1848;
a third at Westhope, Feb. 15th, 1870 ; and a fourth at
Madeley, Dec. 24th, 1890; while a Dunlin in partial
summer dress was seen by Mr. Aplin flying up the Severn,
near Shrewsbury, April 30th, 1888.

Little Stint. Resembles the last but is much smaller. It visits
T. minuta.   Britain on its spring and autumn migrations
                            but no specimen has ever been shot in Shrop-
shire. Mr. Rocke mentions a flock of about twenty, seen on the Teme
in the winter of 1864, which he believed to be of this species.

Curlew Sandpiper. Resembles a very small Curlew. Its
T. subarquata   visits are confined chiefly to the Eastern
Aug.—May.   shores of Great Britain, and occur mostly
                            in August and April. One was shot on
the old Shrewsbury racecourse (near Oxon) in 1836, and
five were killed, out of a flock of about forty, in a turnip
field at Barrow, Broseley, Sept. 9th, 1897. Some of

these were preserved ; one is in the collection of the late
Mr. Beckwith, and another at Willey Park.

Purple Sandpiper. Very rare in Shropshire and has not
T. striata.   occurred lately. It is of a darker
colour, has shorter legs than the other
Sandpipers, and usually frequents rocky coasts.

Knot. Another common shore bird that occasionally wanders
T. canutus.   inland to us. It has occurred on the
Sept.—Apl.   Severn at Buildwas, Cressage, Eyton,
and Uffington—having probably follow-
ed the course of the river from the estuary.

RUFF (Fem. Reeve). This beautiful bird, (of which there
Machetes pugnax.   are some fine specimens in the Shrewsbury Museum), is now rare in Britain
as a resident.   The colouring of the male, with his large
ruff, is subject to such variation that no two birds are
quite alike in plumage.   This ornamentation is only
present in the breeding season.   The Ruff is only recorded
twice in Shropshire: near Melverley in 1861, and Buildwas
1867, both during hard frosts.

Common Sandpiper—*B.* Provincial, Summer Snipe.
Totanus hypoleucus.   The sweet but melancholy pipe of this
Apl. ii.—Sept. iv.   little bird may be heard in summer on
most of our rivers and brooks, especially
near the Welsh border ; it is uttered as the Sandpiper
flits along just under the banks, though the bird has,
besides, a pretty trilling song.   It nests regularly near
Shrewsbury, always close to the water.   On June 18th,
1892, Rev. J. B. Meredith saw a female, near Nesscliff,
fly along the Severn carrying a young one on its back.

Mr. H. H. Hughes once witnessed, on the Teme, the process of mounting—the old bird spread out her tail and wings on the ground, and the little one promptly ran up on to her back, when she immediately flew off across the stream.

Green Sandpiper—*B.* Distinguished from the last by its *T. ochropus.* larger size, shorter legs, and broader dark bands on the centre tail feathers. The Green Sandpiper occurs pretty often in Shropshire, in autumn and winter, and has been seen in summer near Condover and Clun—one shot Aug. 1st, 1891, and another July 13th, 1894. It is now known that, unlike the other Sandpipers, it often lays its eggs in the old nests of such birds as the Thrush and Blackbird, and ignorance of this fact may have caused its eggs to be overlooked. It also sometimes lays on the ground, and in such a situation nested, and reared its young, at Clungunford in 1888. Mrs. Rocke saw the birds in the following summer as well, but no nest was found that year.

Redshank. Distinguished from the other Sandpipers, etc., *T. calidris.* by its long orange-red legs. This bird Mar.—Sept. occurs from time to time, generally on the Severn and Teme. On April 6th, 1892, one was shot near the Dayhouse, Shrewsbury. It is not known to have bred in Shropshire. It is said to be very noisy if its nest is approached.

Greenshank. The legs of this bird are olive-green. It is *T. canescens.* rare in the County but has occurred several times in autumn. A young one was shot near Pontesbury in Sept., 1885, and another near Cressage, Sept. 5th, 1898.

Bar-tailed Godwit. This and the next species have long
Limosa Lapponica. legs, slightly webbed feet, and long
rather upward-curving beaks. The
two are distinguished by their tails being respectively
barred and black as expressed in their names. The
Bar-tailed Godwit has occurred several times in Shropshire on its migrations. One—in the Hawkstone collection—was shot on Hine Heath, Dec., 1849; Mr. T. H.
Thursfield has one, shot at Barrow in 1870; in 1878 two
were shot near Cressage; and on Jan. 29th, 1885, one at
Buildwas.

Black-tailed Godwit. A very rare spring and autumn visitor.
L. Belgica. Rev. W. Houghton saw four in the
spring of 1877, near Preston on the
Weald Moors.

CURLEW—*B.* This large bird, and the much smaller
Numenius arquata. Whimbrel, is easily known by the
curious downward-curving bill. The
Curlew is numerous on our Shropshire moorlands and
breeds regularly on the Longmynd, Whixall Moss, Clun
Forest, and a few other places in the County. Usually
they are very retiring in their habits, but in winter sometimes come down to the lowlands to feed.

Whimbrel. A rare visitor to Shropshire Moors on its spring
N. phæops. and autumn migrations. It has never
bred here. It resembles the Curlew
but is smaller and more slenderly built.

Black Tern. This is the first of a large group of birds—the
Hydrochelidon nigra. Terns, or Sea Swallows, and Gulls—
which are all essentially marine in their
habits, and must be regarded only as stragglers to an

inland county like Shropshire.   Many of them, nevertheless, travel here when there is no gale to drive them over,
and there is no doubt that the Herring and Lesser Black-
backed Gulls often go far from the sea in search of food,
especially in winter.   The Black Tern has occurred
several times in different parts of the County, the most
recent being at Oxon Pool, May, 1871, Rednal, 1873,
Gobowen, July, 1883, and a fine female at Walcot Park
in the spring of 1894.   The tail in this species is only
slightly forked.   The name Black Tern is misleading for,
even in the dark summer dress, the bird is far from black.

Sandwich Tern.   A fine male was found dead near
Sterna Cantiaca.   Shrewsbury in Aug., 1897.   It is now
                          in the possession of Mr. Dumville Lees.

Roseate Tern.   One was shot, about 1830, at Longden-on-
S. Dougalli.   Tern.

Common Tern, or Sea Swallow.   The last name is not
S. fluviatilis.   at all inappropriate, for the Terns all
May—Sept.   have long wings and forked tails and
                            are swift and graceful in their flight.
This Tern is often seen on or near the Severn in autumn
and spring.   The most recent are : one found dead at
Shrewsbury Aug. 21st, 1894, and one shot near Alberbury
March 9th, 1898.   In the summer of 1898 one of these
pretty birds stayed for more than a week on the Severn
below the English Bridge, Shrewsbury.

Arctic Tern.   Closely resembles the last but is slightly
S. macrura.   darker in colour and has legs and beak
                         of a brighter red.   It frequently occurs
in Shropshire, and in May, 1842, immense numbers

appeared on the Severn, in an exhausted condition. One
was shot at Kinnerley, April 12th, 1898.

Little Tern. A much smaller species and occurs less
S. minuta. frequently than the two preceding
May—Sept. Terns. One was shot near Alberbury,
May 3rd, 1898.

Sabine's Gull. A specimen of this small fork-tailed Gull
Xema Sabini. was found dead at Nobold, in the
autumn of 1874; and a male, in full
plumage, was shot at Sandford, near Oswestry, Sept. 8th,
1893, and is now in the possession of Alan Bright, Esq.,
Liverpool.

Little Gull. The smallest of the true Gulls; has been
Larus minutus. obtained at Coalbrookdale, and in 1874
Oct.—Mar. one was shot, out of three, at Atcham.

BLACK-HEADED GULL. Not infrequent during the
L. ridibundus. winter months. The dark-brown (not
black) head is only seen when the bird
is in breeding plumage. Of two obtained near Shrewsbury,
in March 1898, one was in summer, the other in winter
plumage. This bird breeds on marshy flats, and there is
a colony established at Palè, Corwen, many miles from
the sea.

COMMON GULL. The name of this Gull is misleading
L. canus. as it is less common than the Herring
Gull, Lesser black-backed Gull, and
other species. It occurs in Shropshire from time to time,
especially on the Severn near Cressage—generally immature birds in autumn. One was found dead at Bolas,
Feb. 1st, 1890. Most of the Gulls when young have more
or less mottled plumage and are then not easy to identify.

HERRING GULL. Often seen here in winter, especially
L. argentatus. in tempestuous weather, but the immature birds of this and the next species
are so much alike that they cannot be distinguished in
the air. The Herring Gull, however, may be known by
its flesh-coloured feet. When visiting inland places these
Gulls will follow the plough to pick up worms from
the freshly turned earth.

LESSER BLACK-BACKED GULL. Large flocks of
L. fuscus. Gulls, probably of this species, are often
seen flying across the country in autumn
and spring. About forty were counted on Sept. 12th,
1898, and seen by several people about Shrewsbury and
Church Stretton; one shot on the same day out of a
flock at Marton proved to be a young bird of this species.
Several others have been recorded from time to time in
various parts of the County. The feet are yellow.

GREAT BLACK-BACKED GULL. The largest of our
L. marinus. Gulls and only a rare wanderer to Shropshire; one was killed on the Severn
near Cound in Nov., 1861. It has been observed twice
near Knockin—in 1891 and subsequently.

Glaucous Gull. Another large species. Two young birds
L. glaucus. have been obtained; one at Pradoe,
Oct.—Mar. Oswestry, Nov. 1856, shot while feeding
on a dead sheep; and another at
Bomere Pool, December, 1863.

KITTIWAKE. This pretty bird occurs in Shropshire almost
Rissa tridactyla. every winter, generally single specimens,
in an exhausted state—after high winds.

Four occurred in Shrewsbury in one week in Feb., 1899. The Kittiwake has no hind toe. Mr. Harry Shaw had a remarkably beautiful specimen in which *all* the quills had black tips, forming a black border to the wings.

GREAT SKUA. The Skuas are "Pirates" amongst birds, *Stercorarius catarrhactes.* getting their living by attacking Gulls, etc., and making them disgorge their prey, which they catch before it can reach the water. The Great Skua is a large dark bird, only recorded twice in the County. One was caught in the spring of 1879, at Condover, and kept alive for some time; another, the same year, on the Severn above Bridgnorth.

Pomatorhine Skua. This and the two next species have *S. pomatorhinus.* the two centre feathers of the tail much Oct.—Mar. longer than the others. In the Pomatorhine Skua these feathers are twisted vertically and the throat and breast are white. One was killed by flying against the spire of St. Alkmund's Church, Shrewsbury, and there are three in the collection at Hawkstone from Shrewsbury, Shifnal and Baschurch. Another was caught near Downton Castle, March 30th, 1866, and a sixth on the railway at Baschurch, in Oct., 1881.

RICHARDSON'S SKUA. There are two races of this *S. crepidatus.* Skua—a light and a dark one. Specimens obtained in Shropshire have been usually immature, after gales. After the heavy gale of Oct. 14th, 1881, three were found dead and others seen in the County. If disturbed in its breeding grounds, this Skua will fly boldly at the head of any intruder as though about to attack with its beak.

Long-tailed (or Buffon's) Skua. Resembles the lighter
S. parasiticus.   form of the last species but has
Oct.—Mar.   longer tail feathers. There is a young
bird in the Hawkstone collection, shot
near Astley, and one was found dead near Garmston,
October 14th, 1891.

RAZORBILL. All the Gulls and Terns obtain their food
Alca torda.   from the sea, but do not dive. The
Razorbill and all the following birds
except the Petrels, are expert divers, and successfully
pursue fishes in their native element. Most of them fly
heavily and are far more at home in the water than on
land. The Razorbill and Guillemots are so essentially
sea birds that we should hardly expect them to reach
Shropshire. Nevertheless, a Razorbill was caught at
Bromfield in the winter of 1878-9.

GUILLEMOT. Very abundant on the neighbouring coasts
Uria troile.   of North Wales, the Guillemot has
occurred several times in the County, in
autumn and spring. In January and February, 1885,
about fourteen were on the Canal at Ellesmere : several
were shot. Its single large egg is very beautiful and
noted for the extraordinary variety of colour and markings.

BLACK GUILLEMOT. A smaller bird, black, with a
U. grylle.   white patch on the wings. Very rare,
but has visited Shropshire in winter.

Little Auk. Somewhat resembles a Guillemot but is much
Mergulus alle.   shorter and more stoutly built. It is
Oct.—Mar.   only a rare winter visitor to Britain
but a specimen, in the Hawkstone col-

lection, was caught near the Welsh Bridge, Shrewsbury ; and others have been obtained at Shifnal, Acton Scott, Haughmond, Hodnet, Ludlow, and Ellesmere—the last in 1882.

PUFFIN. This comical looking bird, with its large and highly
Fratercula arctica. coloured beak, is common on some of the Welsh coasts, but is a poor flier. Mr. Rocke mentions one found on Corndon Hill. A second occurred at Tasley, Bridgnorth, in Oct. 1887, and another was found dead at Pontesbury, Sept. 24th, 1894.

Great Northern Diver. This is the largest of the three
Colymbus glacialis. Divers and is a very handsome bird.
Oct.—Mar. The Divers are less exclusively marine than the Guillemots, etc., and are often found on inland lochs and meres. An adult female of this species was shot on Ellesmere mere in 1863 and several young birds have been obtained from time to time on the Severn.

Black-throated Diver. Distinguished from the last by its
C. arcticus. smaller size and black throat. An
Oct.—Mar. adult specimen, in the Hawkstone collection, was obtained in 1862, at Gredington, and several young ones at other places, in winter.

RED-THROATED DIVER. The smallest and most
C. septentrionalis. plentiful of the Divers, though rare in Shropshire, where most of the specimens have been young birds. The throat is only red in the breeding season. One was found dead, by a postman, near Market Drayton, in Nov. 1890, a second shot near

Baschurch in Feb. of the same year, and another was
seen on Shrawardine Pool, April 11th, 1897.

GREAT CRESTED GREBE—*B*. The beautiful silvery
Podiceps cristatus. white plumage on the breast of this bird
is well-known on account of its being
used for muffs and collars, etc. In the breeding time this
Grebe has a large brown crest and beard which disappears
in winter. It is resident throughout the year, and breeds
on several of our larger pools and meres; whilst in winter
its numbers are increased by fresh arrivals from the North;
It is then often seen on pools where, however, it is not
known to nest. Of the Great Crested Grebe at Ellesmere,
Mr. Brownlow Tower writes:—" I have counted as many
as twenty-three on the mere at the same time. This is a
rare visitant on many waters, but almost daily to be seen
and watched here. Let us watch one. See his snake-like
head and neck, and how his silvery-white breast gleams
in the morning sun, with his dark brown crest and chestnut
tippet. What dives he takes and how long he remains
under water—when will he come up again ? Never where
you expect; and soon he again vanishes from sight in his
eager chase of some small fish. In the summer time he
may be watched when paying his attentions to his future
wife, to whom he bows and curvets like an accomplished
' beau,' showing off his crest and tippet, as vain of
them as a peacock of his tail."

Red-necked Grebe. A winter visitor to the Eastern
P. griseigena. Counties, but very rare in Shropshire.
Oct.—Mar. It is not so large as the last and has a
smaller black crest and red throat.
The only Shropshire specimens are : one killed on the

Severn near Wroxeter, many years ago, and another close to Shrewsbury, in March, 1888.

Slavonian or Horned Grebe. The head of this bird is, in
P. auritus. summer, ornamented by a long flat
Oct.—Mar. tuft of chestnut coloured feathers on
each cheek, but these are absent in winter. It has occurred in Shropshire several times in winter, the most recent being at Montford Bridge in 1894. One was killed by a boy with a stone in the Quarry, Shrewsbury, in Dec., 1890, and another shot on the Isle pool, Feb. 5th, in the same year.

Eared Grebe. In summer this Grebe has the head and neck black,
P. nigricollis. with a red patch on the ear coverts. Two were
killed on Hanmer Pool in 1864, just outside our County boundary.

DABCHICK, or little Grebe—*B.* An interesting and shy
P. fluviatilis. little bird, common on most of our pools,
where it nests regularly. Like those of the other Grebes its nest is a mass of old rushes and leaves *in* the the water almost submerged. The eggs are white at first but soon become stained by the wet matter of the nest. When alarmed the Dabchick, after hastily covering its eggs with dead leaves, dives silently, and travels under water a long distance before rising again; even then showing little but the head above the surface. Its numbers are increased by fresh arrivals in winter. The Dabchick "does not swim like a Moor-hen, with a jerking movement, but steadily for a short distance and then suddenly dis-appears, making no splash or noise but slipping into the water as if its body was lubricated. It is diving for its food, which consists of water insects, molluscs, and small

fish and worms. As suddenly as it dives, so suddenly does it re-appear, not far from the spot where it was first observed." (*Brownlow Tower*).

Manx Shearwater. In the *Field* of Dec. 11th, 1886, Mr.
Puffinus Anglorum. Beckwith wrote :—" Although said to be so truly aquatic that it seldom resorts to the shore except during the breeding season, the Manx Shearwater sometimes comes inland in autumn and more frequently in wet foggy weather than after storms. In Sept., 1859, two were caught, one on the Severn, in Shrewsbury, the other near Shifnal. In Sept., 1873, another was killed on the Severn at Montford Bridge ; and in Oct., 1877, one joined the poultry at the Hay Farm, near Coalport, where it remained for several days, but was at last killed by a cat. During the early part of Sept., 1882, a small flock of these birds appears to have wandered here. Two were caught near Oswestry, and three others near Ludlow, Ditton Priors, and Cressage, three of them being adult, and two birds of the year. Again in Sept., 1884, an adult bird was picked up near Shrewsbury, and two others—at Aston-on-Clun, and Halston, near Oswestry. Several of these birds that I examined, though in good plumage and condition, were quite unable to fly, and even those that had gained fresh water did not seem able to recover their strength." Another was found by Mr. Benson at Lutwyche, Sept. 9th, 1887.

FORK-TAILED PETREL. A pretty little black bird
Cymochorea leucorrhoa. similar to the next but not quite so deep a black and with a much more forked tail. It has been obtained about ten times in the

Photo by J. Franklin.   Mounted by Wm. Franklin.

PHEASANTS with NEST and TREE-CREEPER.

Photo by Jones & Son, Ludlow.   Specimens at Clungunford.

GREAT BUSTARD.

County, usually after gales, the most recent being at Wheat Hill, Sept. 28th, 1882; Attingham, 1884; at Montford, Sept. 30th, 1886, and on the Teme in October, 1891. This species is often confounded with the next.

STORM PETREL. The "Mother Carey's Chicken" of *Procellaria pelagica.* sailors; has occurred at Wellington, Hawkstone, Ellesmere, Ludlow, and other places after storms. On July 15th, 1886, the weather having been for some time fine and calm, an adult Storm Petrel was brought to Mr. Beckwith that had been caught alive at Homer, near Much Wenlock.

CHAPTER V.

The Wild Birds Protection Acts.

By Watkin Watkins, b.a., Cantab. of the Inner Temple;
Member of the British Ornithologists Union; Barrister-at-Law.

I T is proposed here to give a slight sketch of the general law relating to the protection of wild birds throughout Great Britain and to draw attention to the special regulations now in force in Shropshire.

The general law on the subject of the protection of Wild birds other than game birds is to be found in the Wild Birds Protection Acts, 1880, 1881, 1894 and 1896. The principal Act, the Act of 1880, establishes a close time, between March 1st and August 1st in each year, within which it is unlawful to shoot, or attempt to shoot, Wild birds, or to use any boat for the purpose of shooting, or causing to be shot, any Wild bird, or to use any lime, trap, snare, net, or other instrument, for the purpose of taking any Wild bird, or to expose or offer for sale, or to have in ones control or possession any wild bird. Certain Wild birds are included in the schedule to the Act and these scheduled birds are protected even against owners and occupiers of land and persons licensed by them. Birds not in the schedule are only protected as against persons who are not owners or occupiers of land, or who are not acting with their permission. Orders may be made, on the application of County Councils, to extend or vary the close time prescribed by the Act, and to exempt the whole, or a part of a County, from the operation of the Act.

By the Act of 1894 orders may be made by County Councils to prohibit

(1) the taking or destroying of wild birds' eggs in any year or years in a place or places within the County;

(2) the taking or destroying the eggs of any specified kind of wild birds within the County or parts thereof.

The Act of 1894 also empowered a Secretary of State, on the representation of the Council of an administrative County, to add to the schedule of the Act of 1880, Wild birds not included in the schedule.

By the Act of 1896 orders may be made " for special reasons to be mentioned in the application " prohibiting the taking or killing of particular kinds of Wild birds from August 1st to March 1st, or the taking or killing of all Wild birds. The Act of 1896 first enabled the Council of a County Borough to make orders under the Acts.

The Acts relating to the Protection of Wild Birds now in force in the Isle of Man are more stringent than the corresponding English Acts. The Acts are : The Isle of Man Sea Gull Preservation Act 1867 (32 Vic. 1868) and the Isle of Man Wild Birds Protection Act 1887 (51 Vic. 1887). This latter Act contains provisions, many of which are similar to those in the English Act of 1880, as amended by the Act of 1881. The close time in the Isle of Man is from 1st February to the 1st September, and is therefore longer by two months than the close time under the English Act.

In Jersey, by the Loi sur le port d'Armes the killing of Sea Gulls (mouettes ou mauves) is forbidden throughout the year under the penalty of one livre sterling (*i.e.* about 1s. 3d.) The killing, exposing for sale, selling, buying, carrying or hawking of any sea bird is forbidden from the 1st April to the 1st August under a similar penalty.

Game birds in England within the Game Act 1831 (*i.e.* pheasants, partridges, grouse, heath or moor-game, black game and bustards) have a special close time provided by that Act and cannot fall within the purview of the Wild Birds Protection Acts. As regards the eggs of game birds the powers of County Councils seem more extensive, and the Wild Bird Protection Acts do to some degree trench on those rights of landowners and sportsmen which are recognised by the Game Act of 1831. The Game Act of 1831 protects to some extent the eggs of certain birds, namely, of all birds of game, of swans, wild duck, teal and wigeon; section 24 of that Act makes it an offence for " any person not having the right of killing game upon any land, nor having permission from the person having such right wilfully to take out of the nest or destroy in the nest upon such land " the eggs of the birds named above; this section recognises that the person who has the right of killing game on land has a qualified right of property in the eggs of the birds.

Shropshire has not altered the statutory close time, the only order that has been obtained up to the present is one dated August 4th, 1897. It adds the following birds to the schedule, viz:—treecreeper, spotted flycatcher, pied flycatcher, white wagtail, grey wagtail, yellow wagtail, nuthatch, swallow, martin, sand-martin and swift; and prohibits the taking or destroying of the eggs of the following species of wild birds throughout the County, viz:—crossbill, all owls, nightjar, curlew, kingfisher, goldfinch, nuthatch, all woodpeckers, great crested grebe, nightingale, shrike, white wagtail, grey wagtail, yellow wagtail, spotted flycatcher, pied flycatcher, wild duck, teal and wigeon.

Of the birds which are added to the Schedule in Shropshire the tree-creeper is by no means uncommon in the County

and its nest is so difficult to find that it stands in little need
of protection. The spotted flycatcher is fairly common in the
County and does not need protection. The pied flycatcher is
far less numerous than the preceding species, and during the
breeding season is a very local species, nesting annually in
the English counties on the Welsh border. The white wagtail is a very rare bird and has hardly been known to visit
Shropshire, it may possibly have been included in mistake for
the water or pied wagtail. The grey wagtail is somewhat
locally distributed. The yellow wagtail is not uncommon at
certain seasons of the year throughout the country. The nuthatch is common in certain parts. The swallow, sand-martin
and swift are very common and hardly need protection. The
house-martin has decreased in certain parts of the County
within the last ten years, and if protection is of use, certainly
requires it.

Of the birds whose eggs are protected the crossbill has
occasionally been known to breed in the southern and eastern
counties but its chief breeding haunts are in the north. The
owls are the most useful of all English birds and the protection
of their eggs is only a matter of justice. The nightjar has
become more plentiful of late years. The curlew breeds
regularly in the wilder parts of the County. The kingfisher
was at one time in danger of becoming extinct but is now less
uncommon. The goldfinch is one of the birds that becomes
scarcer every year, partly owing to the fact that cultivation
has everywhere improved and partly owing to the fact that
they are an especial prey of the birdcatcher. The woodpeckers are by no means common, and the protection afforded
to their eggs is most desirable. The great crested grebe is
not uncommon, but it is to be hoped that the protection given
to its eggs by the County Council will lead to the increase

in numbers of this handsome bird on our Shropshire meres. The nightingale is exceedingly scarce in Shropshire and perhaps no protection will ever make it abundant in the County.   The red-backed shrike breeds in Shropshire but by no means plentifully, its nests are hard to find but eagerly sought after, and the protection given to its eggs is certainly needed.   The eggs of the wild duck, teal, and wigeon are duly protected by the Game Act of 1831 and it is doubtful if the inclusion of these birds in Orders made under the Wild Birds Protection Acts is expedient.

There are some birds which are not protected by the order of the Shropshire County Council and which stand in need of additional protection, such as the heron, the jay, the hawfinch and the kestrel.   The heron is a welcome addition to the landscape ; it wanders far from its home and is often shot by owners of land, especially those who are interested in fishing.   The jay is in the gamekeeper's black book and no doubt takes eggs during two months of the year, but is harmless during the other ten, and is such a beautiful bird to look at, that its peccadilloes might well be pardoned for the sake of its appearance.   The hawfinch finds an enemy in the gardener, but it is so local that it ought to be protected. The kestrel is, after the lapwing, the bird that does most service to the farmer, for it is a most excellent mouse-trap. The lapwing, the farmer's best friend, might well find some protection for its eggs and the example set in Scotland might well be followed in England, namely, to protect the eggs after the 15th April, and thus, while allowing the first clutches of eggs to be taken, to ensure the second clutch being unmolested. Protection is also given in many parts of Great Britain to such birds of prey as the kite, buzzard, raven, harrier, peregrine falcon and sparrowhawk, and some or all of these birds might well be protected in Shropshire. It has been suggested that the great increase in recent years of sparrows and of wood-pigeons, has been owing to the killing down of their enemies, the raptores. It may be that by protecting such birds of prey as those just mentioned, their ravages may be checked ; whether this be so or not, it is hardly likely that birds of game will suffer much, or that anything but good will result from the encouragement given to nature's scavengers and vermin destroyers.

CHAPTER VI.

REPTILES.

WHAT is a Reptile? The word means literally anything that creeps, but in scientific phrase is restricted
in its application to the creatures known as Snakes, Lizards,
and Tortoises. It is also sometimes applied to Frogs, Toads,
and Newts, but, as will be shown later on, those animals
differ widely from the true Reptiles in many respects. Without going into anatomical details, it will be sufficient to state
a few of the most marked peculiarities of the Reptiles. As
distinguished from Mammals and Birds we notice first that
they feel cold to the touch, and the blood is cold; this
may be due, in some degree, to the circulation being
imperfect, the heart having only three instead of four
chambers. All Reptiles breath by lungs, and *at no period of
their lives* are provided with gills. The young undergo no
metamorphosis, and are produced from eggs; these are not
hard-shelled, however, but are enclosed in a leathery envelope.
Sometimes this envelope ruptures and liberates the young
before the egg is extruded from the body, and the animal
is then said to be ovo-viviparous. The Viper, the Blind
Worm and the Common Lizard are British examples of
this phenomenon, and it is possible that our cold climate
has induced this state of things, as it would be advan-
tageous to the species for the young to be born alive,
rather than run the risk of perishing if exposed to the
elements. In this way we can account for the fact that three
out of the six reptiles found in Britain are ovo-viviparous

though in hot countries nearly all are oviparous. Snakes and Lizards cast their skins periodically—at least once a year—and generally make a meal of the old coat. As in other creatures that change their skins, the colours of the new coat are brighter than those of the old, and at this period the animal is very susceptible to cold. The skin in Snakes is covered with what appear to be scales, but really each of these is a flattened fold, or pocket, in the true skin, and the outer covering of each will be found intact, forming part of the cast "slough." The truth of this is evident if the skin of a snake and fish be rubbed ; it will be found impossible to remove the projections on the snake's skin, while the scales of the fish (being only outside appendages to the skin), rub off easily. The teeth are generally conical and simple, and fixed to the jaws by a bony cement instead of being inserted in sockets. The eyes are provided with eyelids in the Lizards, but Snakes have an immovable covering like the cover-glass of a watch. Snakes have no external traces of limbs, but there is evidence that they are descended from an ancestral type that had limbs, as some of them possess internal rudiments of legs. Reptiles, being cold-blooded, are of a sluggish disposition, and except when roused by hunger or other passion, display little vitality. Cold weather makes them sleepy, and they then retire into holes, and coil up to keep themselves as warm as possible : frequently several are found together rolled up into a ball. In England they usually come abroad for only the six warmer months of the year, sleeping through the remainder of the time. The vertebral column is always long, and, in some of the Snakes, contains a very large number of vertebræ. The feet, when present, are usually provided with five toes, without claws. Most writers state that Lizards have the power of reproducing a lost tail, or limb, but there is no doubt that some of these accounts have been exaggerated. The tongue in Reptiles is peculiar; in most species it is long, flexible, and notched, or forked, at the end, and the root is near the front of the mouth instead of being close to the throat, so that when the tongue is exserted it projects a long way beyond the lips. Snakes have a sheath into which the tongue can be withdrawn, but this is absent in the Lizards. To the rustic mind, a Snake with its forked tongue curling about in the air, is a fearsome object, and the usual exclamation is—Oh, look at its sting! The tongue is, of course, perfectly harmless, and it cannot be too widely known that, in England, none of the Reptiles are venomous—that is to say, possessed of poison-fangs—except the Viper, and even in this case the bite is very rarely fatal, although it causes considerable pain and swelling in the affected part. It is much to be regretted that the word "reptile" has acquired such a sinister meaning, that it is used only in reference to creatures that should be avoided and treated with loathing. If people could only get rid of their prejudices they would find a great deal to admire in the sinuous grace and beautiful colouring of the common Snake, and the agile movements of the pretty little Lizard. As a rule Reptiles prefer warm to damp places, and are found in hot countries in great numbers and variety; Britain is too cold to suit them well, and we have, consequently, very few species, and those of small size. Only six are recognised as British species; five are known to occur in Shropshire.

COMMON LIZARD. Provincial name, Harriman. This
Lacerta vivipara.　　　　pretty and agile little creature, the
　　　　　　　　　　　smallest of our Lizards, is fairly common on heathy uplands in Shropshire. Mr. J. Steele

Elliott reports it as common in the Wyre Forest, and Mr. Martin J. Harding says he has seen it frequently along the old Potteries Railway line, near Shrewsbury. Like other reptiles, the Lizard is fond of basking in the sun, and when thus engaged will often allow itself to be approached very closely. As a rule, however, it is very timid, and, on the least alarm, darts off to its lurking place in a crevice between stones, etc. The most marked peculiarity of this species is that the eggs are hatched inside the parent's body, so that the young are apparently born alive as in the Mammals—hence it is called the Viviparous Lizard. The resemblance to mammalian generation is, however, only superficial, for each embryo is contained in a separate egg with the enclosing membranes, exactly as in those reptiles that deposit eggs in the ordinary way. The female when breeding lies in the sun as much as possible, to facilitate the hatching of the eggs, and at such times is very unwilling to be disturbed. The young—generally three or four in number—are almost black, and have the use of their legs as soon as they are born. Just at first they keep with the mother, but, so far as is known, she does not feed them, and driven by the feeling of hunger, they soon learn to catch insects, etc., for themselves. The tongue and mouth are chiefly used in capturing their prey, the limbs being used only for progression. All the movements of the Lizard are graceful and lively, and if an insect comes within reach it darts upon and swallows it with the rapidity of lightning. The food is entirely of an animal nature, and consists principally of insects, worms, and slugs. It is said to take readily to the water, and to swim rapidly, and, in comparison with the Sand Lizard, to exhibit a decided preference for high ground. The Common Lizard usually lives on or near to the ground, but the writer has seen it climb the stem of a big gorse bush to reach a sunny place to bask in. If kept in a fern-case, the Lizard makes a pretty and interesting pet, and soon loses its shyness. To capture one it is best to employ a butterfly-net, and great care has to be used in handling it, for, if seized by the tail, that member immediately breaks off in the hands of the captor, and commences a series of lively contortions, while the Lizard makes its escape. The severed tail retains its power of independent motion for several minutes. This extraordinary self-mutilation is performed voluntarily by the reptile, and is effected by the sudden violent contraction of certain muscles in the tail. A fresh tail is soon formed in place of the lost one, but it is slightly smaller than the original, and shows a mark at the point of severance. This phenomenon occurs in all Lizards, and it is said that they will reproduce a lost limb in the same way. Occasionally Lizards have been found with two tails; in these cases the original tail had been only partially severed, but a second one had grown from the wounded spot. The colour of the Common Lizard is greenish-brown, with black lines along the whole length of the back, and the under-parts are yellow with black spots. The head is flatter, and the scales on the body larger than in the Sand Lizard, while the whole animal is smaller, rarely exceeding six inches in length.

SAND LIZARD. Less common than the preceding, this
L. agilis. species is found more frequently on low-lying ground and on sandy heaths.

In Shropshire it has been found on the Weald Moors, by Eyton, and more recently near Admaston, by Rev. W. K. Wyley, and at Bromfield, by Mr. Henry Gray. The last‑named gentleman reports a very curious incident witnessed there not long ago :—a Sand Lizard emerged in front of the observer from a bank close to a field path, and ran across the path; it was followed almost immediately by a Mole, which tracked out the Lizard, apparently by scent, following it up through all its turns and windings till eventually it came up with it, seized it by the body, and bore it back in triumph to the same hole from which it had previously issued! If it were not stated on such good authority, it would seem unlikely that the Mole could overtake the Lizard, and it is an interesting matter for consideration, how the Mole, while still underground, could be aware of the presence of the Lizard overhead. It is stated in books that the mole will eat lizards, but the writer had not previously met with an actual instance. The Sand Lizard is less agile in its movements than the Common Lizard, and this, of course, would be in the Mole's favour in a trial of speed. Several species of Birds and other Mammals, besides the Mole, eat lizards, and it is a well‑known fact that some of the larger tropical lizards, such as the Iguana, are cooked and eaten by the natives, and are said to be very toothsome dainties. The Sand Lizard, unlike the common species, is ovipar‑ous, laying about a dozen eggs in a slight hollow scraped in the ground, covering them with a thin layer of sand, and leaving them there to be hatched by the heat of the sun. The food and general habits are similar in both species. The Sand Lizard is distinguished from the

Common Lizard by its heavier build, larger size, more sluggish movements, oviparous habit, by the presence of minute teeth on the palate and granules over the eyes (absent in the common species), and by the more numerous, but smaller scales on the body. In colour it varies greatly, and, according to some authors, these variations coincide with the habitat, so that the animal is brownish when it occurs on sandy heaths; greenish when it lives in grassy places. Be that as it may, the ground colour of the back is always some shade of brown or green, while the sides are greenish in the male, brownish in the female, with a varying number of white spots ranged usually in three longitudinal rows. The under parts are white with small black spots. Length, about eight to nine inches. Like other reptiles it passes the winter in a torpid state. It is said to refuse food in confinement, and is not so readily tamed as the preceding species.

BLIND-WORM, or SLOW-WORM. Truly appearances
Anguis fragilis.   are deceptive, and accordingly almost
                                      with one accord this harmless little
creature is dubbed a Snake, and very venomous, though it is neither the one nor the other! The names by which it is generally known—Blind-worm and Slow-worm—are equally incorrect: the animal is not blind but has small bright eyes; it is not particularly slow, for it can wriggle along the ground rapidly enough; lastly, it is not a worm but a lizard! Yes,—a lizard, but without legs; under the skin, however, it has rudiments of limbs, and if we look carefully we can see certain places on the body where the scales stand out as if covering a knob, and these indicate the position which the hind-limbs would

occupy if they were present. The presence of movable eyelids, the solid lower jaw, the notched (not forked), tongue, and other anatomical characters, all indicate that the animal is a lizard and not a snake. It gets its specific title of *fragilis* from the habit of breaking off its tail when alarmed, a habit common to the lizard tribe, but, as far as the writer has observed, this habit is much less marked than in the other lizards, and he believes that the Blind-worm never does this unless startled or seized very suddenly. The creature lives mostly in crevices in the ground, and along hedge-banks, and, like other reptiles, emerges from its retreat when the weather is warm, to bask in the sun. It is then so supremely happy that it becomes oblivious of all else but the pleasant warmth, and can be caught easily by anyone who approaches it quietly. If it is desired to secure it without the loss of its tail, it is best to slip a butterfly-net over it *not too suddenly*. When handled it will bite savagely at the fingers, but is utterly powerless to inflict harm, its teeth being too minute to penetrate the skin, and of course it has no poison-fangs. It feeds chiefly upon snails and worms, poising its head in the air preparatory to pouncing on them and seizing them in its jaws. When hungry the Blind-worm keeps licking the outside of its lips with the notched tongue, and this, to the rustic mind, is evidence of its possessing a venomous sting! As in the allied reptiles, the skin or slough is cast entire, periodically. The Blind-worm is plentiful throughout Shropshire, in such situations as are suited to its habits, and this too in spite of the numbers killed every year by the mis-directed zeal of ignorant persons. It is—like the Common Lizard—ovo-viviparous, producing some eight or ten young, about July. In colour it is metallic-grey (sometimes with a coppery tinge), along the back and sides, paler underneath. The average size is from ten to twelve inches, though occasionally found over fifteen inches in length. The entire body is covered with small scales, and is of almost uniform thickness throughout; the tail tapers slightly and is not very sharply pointed.

COMMON, or RINGED SNAKE. This is the most *Tropidonotus natrix.* abundant of the Reptiles in Shropshire, and is found pretty generally throughout the County. It is also the largest British Reptile, occasionally reaching a length of five feet; such a size is, however, quite exceptional, and it is more often found measuring between thirty and thirty-six inches. The colouring is really beautiful, a combination of various shades of green and yellow. It gets its name of 'Ringed Snake' from the patches of bright primrose-yellow on the neck, forming an almost complete ring around it. It is rarely found far from water, and "may often be surprised, coiled up in sunny weather, with its head out, enjoying the luxury of a bath. It will dive after water-newts, especially when rather hungry, bringing them to shore in its mouth, and devouring them upon dry land" *(M. C. Cooke).* It is also said to capture fish in the same manner, and in the *Royal Natural History*, the Ringed Snake is represented as swimming with the head above water and bearing off in its mouth a small fish. There is abundant evidence that the Snake prefers Frogs to any other kind of food, though it also feeds freely upon mice, eggs, and young birds. The snake's mode of feeding is peculiar: the lower jaw is not articulated direct

Photo by J. Franklin.   Mounted by W. Franklin.

OYSTER-CATCHERS.

GROUP OF SKUAS, ETC.

1, 2, 3. BUFFON'S SKUA (F., M., Y.)
4, 5, 6. RICHARDSON'S SKUA (Y., F., M.)
7, 14. MANX SHEARWATER (F., M.)
8, 11, 15. POMATORHINE SKUA (M., Y., F.)
9, 10. GREAT SHEARWATER (F., M.)
12. BULWER'S PETREL.
13. FULMAR PETREL.
16, 17. COMMON OR GREAT SKUA (F., M.)
18, 19. WILSON'S PETREL (M., F.)
20. STORM PETREL.
21, 22. FORK-TAILED PETREL (F., M.)

to the skull, but to an intermediate bone called the quadrate ; the right and left jaws are also two separate bones. The whole of these parts are connected together by elastic ligaments; thus it happens that the snake is able to swallow animals entire that are many times larger than the normal size of its own mouth. It does not masticate its food, and the teeth are only of use in holding its prey; hence the animals are swallowed whole and *alive*, and instances are recorded of frogs being taken from the stomach of the Snake and hopping off, like Jonah, none the worse for their unpleasant experiences! When a Snake seizes a frog it is generally by one of the hind legs, and the frog soon gives up all attempts to escape; the Snake then by a continuous movement of its jaws works its way up the leg till it reaches the body ; the other hind-leg is now forced forward alongside of the body, and as that is gradually drawn into the now greatly distended maw of the Snake, all three feet are grouped together round the frog's head as they slowly disappear. After a good meal the Snake shows a strong inclination to gape ! If the Snake happens to seize a frog by the middle the procedure is somewhat different —it then turns it round in its mouth and swallows it, head first. It is said that the gaze of a Snake exercises a kind of fascination upon small animals and birds, so that they are unable to escape, waiting in a kind of mesmeric trance till the Snake chooses to devour them. This rather ghastly spectacle may be witnessed in the Snake House at the Zoological Gardens, whenever the larger foreign snakes are fed, but we are not sure to what extent it takes place with our native snakes when at large. Frogs are said to utter a peculiar croak of

L

alarm, never heard at any other time, and birds and mice sit still, trembling with fear. A few years ago the keeper at Betton killed a Snake; noticing that one part of its body was greatly enlarged, he opened it, and found three Water Voles; the one that had been swallowed first was partially digested, the other two were quite fresh. If irritated, the Snake utters a hissing sound and at the same time emits a strong pungent odour, which clings for a long time to the clothes of anyone who handles it. As mentioned at the commencement of this chapter, the scale-like covering of the Snake consists of a series of pocket-shaped folds of true skin. The slough is cast several times a year. Just before the change the colour alters to a dull grey, and the Snake seems restless and uneasy; the skin then splits open on the neck, the edges curling outwards; the Snake wriggles gradually out of the old skin, and in the process turns it completely inside out. If the slough is examined it will be found to have every detail impressed upon it, every ' scale ' perfect, and even the covering of the eyes intact, but, being reversed, these appear concave instead of convex. Snakes have not movable eyelids, but possess a glassy covering to the eyes, formed by the fusion of the upper and lower eyelids. It is this that imparts such a peculiar aspect to the head—the unwinking stare so characteristic of these reptiles. The Common Snake is oviparous, laying some twenty or more eggs, in a dung-heap, or other warm situation; they are gelatinous, with a tough envelope and, although laid singly, usually adhere together in a bunch. As soon as they are laid they absorb moisture and swell till they are rather larger than a sparrow's egg. The length of time they take to develop varies according to the temperature, and they have been known to remain dormant through the winter; not developing till the following spring. Generally, however, they hatch in a few weeks. The young are at first almost black. The mother takes no interest in her offspring, leaving them to shift for themselves. Probably they feed on grubs, etc., that they would find plentiful in manure, thriving on these till they grow big enough to capture larger animals. If the scales on the back of this Snake are examined, it will be noticed that each has a raised ridge or keel running along its centre—a feature which is absent in the rare Smooth Snake to be next described. The tail of the Common Snake is most characteristic, being very long, and tapering off gradually to a fine whip-like extremity. This alone renders it easy to distinguish it from either the Viper or Smooth Snake, both of which have comparatively blunt tails.

Smooth Snake. This species has not been taken in Shropshire, but
Colonella Lævis. Mr. W. S. Buddicom saw a Snake at Ticklerton, about 1880, which he feels sure was neither a Viper nor a Common Snake, and upon reading the description of the Smooth Snake, concluded that it must have been one of that species. Under the circumstances it has been thought advisable to give a description of the Smooth Snake sufficient to enable anyone to identify it, and it is requested that anyone fortunate enough to find one will communicate the fact, or send the specimen, to the author. Colour :—greenish-brown above, with two longitudinal rows of small black spots: lower parts lighter, but much obscured by black markings; head small and rounded, neck very little narrower; scales numerous, small, and smooth *(i.e. not keeled)* ; tail thick-set and not ending in nearly so long and sharp a point as in the Common Snake; size small, rarely reaching two feet in length. It is oviparous, and said to be rather savage, but the bite is quite harmless.

VIPER, or ADDER. Provincial name, Etther. This, our
Pelias berus.   only poisonous reptile, is easily dis-
                                   tinguished from the Snake by several
obvious characters. First, the head is of a different
shape, being wider across the muzzle, and the scale-like
plates on the top are more numerous and smaller than in
the Snake. Second, there is a dark mark, shaped like a
V, on the top of the head, and a wide dark zig-zag line
along the back. Thirdly, the shape of the tail; this is
thick almost to the end, and tapers off abruptly to a
blunt tip. Fourth, the Viper is usually smaller (rarely
two feet in length), and duller in colour than the Snake.
In the field it will be found that the two last points are
the most useful to remember, as the Viper is a timid
creature; if a glimpse of it is obtained it is generally
in full retreat, and probably only the tail can be seen as
it disappears. If this is blunt and dull-coloured it must
be either a Viper or a Smooth Snake, and the chances
are a thousand to one that it is the former—it cannot be
a Common Snake. The nature of the place is also some
guide for—as before mentioned—the Snake loves damp
places, near water, but the Viper prefers dry and warm
situations. In Shropshire the Viper is not nearly so
numerous as the Snake. It is entirely absent from the
immediate neighbourhood of Shrewsbury, the nearest
places where it occurs being Nesscliff and Pim Hills. It
is fairly numerous in the neighbourhood of Oswestry
and Ellesmere, on Rudge Heath and Whixall Moss
in the North; and on Titterstone Clee Hill, and in the
Forest of Wyre in the South. The Viper varies a good
deal in its ground colour, but is generally either of a
dull brownish-grey, or a coppery-red hue: according to

Rev. J. T. Lea, who dissected a large number of specimens, found near his residence (Far Forest Vicarage), in
the Wyre Forest, the difference coincides with the sex ;
the grey ones being males, and the coppery ones females.
At a meeting of the Caradoc and Severn Valley Field
Club, held January 18th, 1897, Mr. Harold Peake, of
Ellesmere, sent for exhibition three very young Vipers
(taken near that town), each of which had two small
hind legs : these are believed to be unique, and are
highly interesting, since (according to the doctrine of
evolution), rudimentary organs found in the young, indicate that the ancestral type possessed those organs in
a more perfect form ; that is to say, these rudimentary
legs indicate that the Viper has descended from ancestors
which possessed legs. This accords with the belief of
most modern biologists that snakes are descended from
lizards. The poison apparatus of the Viper is similar
to that of others of the venomous tribe of Snakes. It
consists of two poison fangs only, found in the upper
jaw, and these are very different from ordinary teeth. In
most poisonous snakes there are ordinary teeth in the
lower jaw, but none in the upper. The poison fangs,
unlike ordinary teeth, are *hollow*, and *hinged at the base.*
When not in use they each lie in a groove along the
upper jaw, with their points directed backwards, but at
the will of the animal they can be erected by a small
muscle, and then stand at right angles to the jaw, with
their points directed downward. When a Viper intends
to strike, it gathers itself together in a heap on the
ground, with the neck raised in the centre and the head
drawn back; it then suddenly uncoils itself, like a spring,
throwing the body forward with extreme rapidity, though

the tail never leaves the ground. At the moment when it springs, the mouth is widely distended and the fangs standing out from the jaw ready to plunge into the enemy. This is effected by striking downward with the upper jaw, while the entrance of the poison fangs into the wound is assisted by the closing upwards of the lower jaw. There is a poison-bag at the base of each fang, and the pressure causes some of the venom to pass through the tubular fang into the wound. Sometimes the Viper bites without first coiling and springing. The wound shows as two small punctures in the skin of the victim. After biting, the Viper immediately withdraws the fangs and coils ready to strike again if necessary; but that rarely happens. Mr. Dumville Lees writes that he has seen Vipers several times on the Moelydd, near Oswestry. One of them bit a dog of his in the mouth; the head swelled very much, and he was so ill he could hardly be got home; ammonia was applied externally and internally and by next morning he was nearly well. Rev. J. T. Lea says that a dog of his that was bitten on the leg, did not recover for six weeks; the remedy used being ash-buds and skim milk. That the Viper does not *always* use its fangs when biting is proved by the following curious incident:—Mr. Ramsbotham, of Meole Hall, Shrewsbury, has a Viper (taken in 1889), which was put into an empty bottle and there left for nearly twenty-four hours; upon the bottle being filled with spirit, a full-grown Lizard *(Lacerta vivipara)*, crawled out of the Viper's mouth! This proves three things:—(1) that a Viper sometimes eats lizards, though its food usually consists of mice and frogs, (2) that it sometimes swallows its prey without using its poison fangs; and (3) that a lizard can exist for nearly twenty-four hours in a snake's gullet. This last point has an important bearing upon an oft-disputed matter—does the Viper swallow its own young? It has been repeatedly stated that if a Viper is surprised with her brood of young ones around her, she will open her mouth wide and the little ones promptly run into it, and take refuge in her stomach till all fear of danger be past! The statement is so strange that it has been received with incredulity, and it is only after reading the independent testimony of many witnesses of undeniable veracity, whose accounts all agree as to the main fact, that we are inclined to think that it may possibly be true.* The strongest objection that could be urged against it is that, if the young were retained in the stomach or gullet of the parent for any length of time, they would be killed, or at any rate injured, by the action of the digestive fluids. This objection is nullified, however, by the instance given above, where a lizard remained *alive* for twenty-four hours in a Viper's stomach; and in all the cases observed, the young Vipers swallowed by the parent have been found to be quite uninjured. Perhaps the digestive fluids of Reptiles act less rapidly upon the skins of reptiles than upon other animal matter—but this is only a conjecture. Mr. R. de G. Benson states that some members of his family once observed, at Pulverbatch, a *Snake* on a hedge-bank, with a brood of young ones who took refuge, when alarmed, in her mouth. The author thinks this must have been a Viper, as the Snake has never been known to act in this way, and, being oviparous, does not take

* Mr. J. Steele Elliott writes The *Field* has offered £5 to anyone who will send them a Viper with young in its stomach; no one has ever obtained the reward.

any interest in her progeny. He mentions this, however, because it appears to be a 'local' instance, and is therefore worth quoting, and if the reptile really was a Snake, the fact is doubly strange.   There are cases on record of fatal results following the bite of a Viper, in this country, though they are rare; the bite, however, causes severe pain and sickness, and it is therefore well to know how to remedy or mitigate its effects.   The first thing to do is to suck vigorously at the wound (which can be seen as two tiny blue spots on the skin), to draw out the venom as much as possible [no one need fear to do this as the venom is not a poison, even if swallowed.] Next, apply warm sweet oil, or liquid ammonia to it as an outside dressing, and take dilute liquid ammonia (Sal volatile), internally.   The virulence of the poison depends partly on the condition of the Viper at the time; if it has not used its fangs for some time the venom bags will be quite full and the quantity of poison injected into the wound will be considerable, but if it has used its fangs not long before, the bags will be partially emptied and consequently less venom will be injected.   Hot weather increases the rapidity of the poison's action.   If the Viper's mouth be examined a second rudimentary pair of poison fangs will be found at the rear of the front pair, and should one of these be lost by accident, it is replaced by the development of the smaller one behind it.   The Viper will climb trees to rob birds' nests; an instance is related by Rev. J. T. Lea *(Shropsh. Archæol. Soc. Trans.,* 1889*),* in which a boy climbing up to a nest in a tree near the Wyre Forest, found it occupied by one of these reptiles, which saluted him with a hiss, and would have bitten him had he not beat a hasty retreat.

CHAPTER VII.

AMPHIBIANS.

THIS order of animals, while presenting certain features —such, for instance, as the cold blood, oval corpuscles, and three-chambered heart,—common to the Reptiles, has several characteristics which are very distinctive. Leaving out of consideration a few abnormal foreign species, it may be well to sum up here those features which are found in all British Amphibians, and to give under the heading of each species only those peculiar to itself. The title " Amphibia " is better than the obsolete " Batrachia," because it calls attention to the fact that all the animals grouped under it are amphibious in the sense that they spend part of their existence in the water, when they breathe by gills; and part on land, when they breathe by lungs. They all undergo metamorphosis: the young or larval forms are called tadpoles, and in many respects present strong similarities to fishes, breathing by gills which are at first external, afterwards internal, and having a flattened fin-like tail which, however, is not supported on bony rays. The tadpoles are in most cases herbivorous, feeding chiefly on confervoid algæ, but the adults are always carnivorous. It is well-known that in all animals that are herbivorous, the intestine is long, whilst in those that are carnivorous it is short and simple. Strange to say the Amphibia begin life with the intestine of the herbivorous character, and as the metamorphosis progresses this changes gradually to the shorter simpler form found in the adult carnivorous animal. The skin in Amphibians is soft, and devoid of scales: the outer skin *(epidermis)*, is colourless, and is cast—usually entire,—from time to time. The skin plays an important part in the economy of these animals, for through it, and the numerous glands on its surface, they absorb the large amount of water that is indispensable to their existence. They never drink through the mouth. The skin also acts as an additional lung, or gill, in the function of respiration. Unlike true reptiles the Amphibia are never found in hot dry situations. It is true that Frogs may sometimes be found sitting in the sun and enjoying the warmth, but this is always close to water, to which they retreat as soon as the skin begins to parch. As a general rule they do not come abroad by day at all, unless the weather is wet; and prolonged drought is fatal to them. They bury themselves in the earth and remain torpid through the winter. There are several peculiarities in the skeletons of Amphibians. The number of vertebræ is smaller than in the Reptiles—in Frogs and Toads there are only seven or eight—and *there are no ribs*. In all other animals that breathe by means of lungs, the ribs play an important part in the act of drawing in and expelling the breath. The Amphibians, having no ribs, are unable to breathe in the ordinary way, and take in air in great gulps, closing the lips tightly each time in order to retain it. So impossible is it for them to breathe without thus holding in the air, that they soon die if the mouth is wedged open. Teeth, if present, are generally found in the roof of the mouth; they are simple and pointed, and are used for prehension only; never for mastication. The tongue is very peculiar; it is soft and, in most cases, slightly notched at the point, and the root is fixed only just inside the lower jaw. When at rest the point is directed towards the animal's throat, while, when it wishes to seize an insect, the tongue is thrown forward with extreme rapidity, turning on its root as on a hinge, and returning in the same way bearing the prey adhering to it. The eggs of Amphibians differ from those of Reptiles in being devoid of the investing membranes round the embryo, known as amnion and allantois. In this, and in the fact that they are fecundated after extrusion, they resemble the ova of Fishes. In brief, it may be said that the Amphibia are intermediate in character between the Reptiles and Fishes—in the adult state they most nearly resemble the former; in the tadpole state, the latter. Many years ago the writer made a special study of the development of Frogs, Toads, and Newts, and as he has found that the subject is one about which little seems to be generally known, he trusts he may be excused for describing it at some length. All the Amphibia are terrestrial in the adult state, but they all deposit their eggs in water, generally such as is of a stagnant character. There is no union of the sexes, but, in Frogs and Toads—as in Fishes—the milt is deposited upon the eggs at the same time that they are extruded into the water. In Newts, the packets of spermatozoa, as they are deposited, are conveyed by the females into their own reproductive chambers. While this is going on Frogs and Toads keep up a loud and incessant croaking and, to an onlooker, it will be a matter of surprise to find that there are so many of these creatures existing in his own neighbourhood. On two occasions the writer has seen toads assemble in crowds that might be estimated at many hundreds, if not thousands, although, in the same district, at any other time, it would be hard to find a dozen individuals. On examination it will be found that this vast assemblage consists almost entirely of pairs—each female accompanied by a male—and that the latter is in most cases smaller than his mate. In all Amphibia the eggs are enclosed in a gelatinous sheath which, as soon as they are laid, rapidly absorbs water and swells up till it forms a thick transparent layer, through which the small dark spherical yolk can be plainly seen. The form of this gelatinous envelope varies with the genus. In the Toad (*Bufo*), it is long and rope-like, the eggs being disposed in a zig-zag manner throughout its length, and the whole hangs in festoons from the water-weeds. In the Frog (*Rana*), each egg is in a separate globular sheath, but the whole mass of eggs adheres together in an irregular heap. In the Newt (*Molge*), each egg is deposited separately and wrapped by the parent in a leaf of a water-plant, held in a coil around it by the glutinous nature of the envelope. Very rarely, the writer has known the Newt to deposit its eggs on the clay at the edge of a pond and spread the clay over them, but this was because there were no suitable weeds in the pond. The yolk in the eggs that are laid in the early spring is black, but in those deposited later on in the year it is of a lighter colour. It has been surmised that this difference is due to the fact that the early eggs need all the warmth they can collect, and *black* absorbs more heat-rays than any other colour. There is also a difference in the buoyancy of the spawn at the two periods—that laid in the cold season is heavy and remains on the ground below the water, where it is beyond the reach of night-frosts; later on the spawn floats. Temperature and sunshine are important factors in the development of the spawn, as also in the rate of growth of the tadpoles after hatching, and if either eggs or tadpoles are kept cold and in a dark place, they may be retarded almost indefinitely. Under favourable conditions the eggs hatch in about twenty-one days, but they often remain dormant for many weeks. The tadpoles in their early days are all alike, though in subsequent stages they exhibit marked differences according to their Frog or Newt parentage. All have a large round head, no body to speak of, but a flattened tapering tail, which is used as a propeller. Under the chin is a pair of processes forming a sucker by means of which they anchor themselves to any convenient object, and they generally associate in clusters while very young. When first hatched, the tadpole is still in a semi-embryonic condition having neither eyes, ears, nor breathing apparatus, respiration being effected chiefly through the skin; in about four days the organs just named make their appearance, while on each side of the neck there may be seen a small knob from which there soon issues a pair of branching gills forming a frill (larger in Newt- than in Frog-tadpoles). These, as well as the whole of the tail, are semi-transparent, and if the tadpole be placed in the microscope, the action of the heart (which consists of two chambers only), can be plainly seen, as well as the blood with the large oval corpuscles coursing through the gills, and the capillaries in the tail. These vessels all arise from a large artery running along the centre of the tail, and return their contents into a vein parallel to the artery. The gills are covered on the outside with cilia (minute vibrating hairs), which by their action produce currents in the surrounding water and so bring up a constant supply of fresh water to aerate the blood coursing through them. At first the tadpole is nourished by the remains of the yolk-sac inside its body, but that is quickly absorbed and it then begins to feed voraciously on confervoid algæ, etc. The lips are soft at first; later on they are produced into a kind of beak, provided with a horny sheath, of great use in dividing the food. The gills reach their maximum size before the production of any limbs, but about the time when these appear they diminish in size externally, and gradually disappear. Meanwhile a fresh and more complex pair of gills are being formed inside, in pockets, or chambers, one on each side of the mouth. These gills are arranged in tufts on four cartilaginous arches, and number over a hundred on each side. When they are fully formed the external gills entirely vanish. The water now passes in through the mouth over the new gills and out through a *single* tube opening on the lower face, or left side of the body. These gills resemble those of fishes in their function of respiration; they differ from fishes' gills in having a branched instead of laminated form, in having a single outlet instead of one each side, and in having no bony gill-covers. The gills in newt-tadpoles continue *external* till they attain the adult quadrupedal form and leave the water. Soon after the external gills are developed, the tadpole grows a pair of limbs. [Hind-limbs if Frog or Toad; front-limbs if a Newt.] These appear first as knobs on each side of the body, and steadily develop till they assume the shape of the legs in the adult. The second pair of limbs is formed subsequently in the same way, while simultaneously the tail gradually shortens [except in Newts], till it finally becomes a mere rudiment or stump, and the animal is a perfect Frog, or Toad, as the case may be. Not *quite* though—Before finally leaving its nursery the little creature has got to become an air-breather, and accordingly during the process of its metamorphosis it has developed inside the body true lungs, which, although present before in a rudimentary solid condition, only now open out and increase in size ready to come into play when it leaves its native pond or ditch, and starts afresh as a terrestrial and carnivorous animal. The

gills being no longer required now shrink up and disappear.
Simultaneously with the development of the lungs and
atrophy of the gills, a remarkable change takes place in the
circulatory system; as long as the animal was a gill-breather,
it had a heart consisting of two chambers only—one auricle
and one ventricle—as in Fishes; as soon as it becomes a
lung-breather the heart develops a second auricle, and then
consists of three chambers, as in most reptiles. Truly we
here see a wonderful transformation—or rather series of
transformations—quite as marvellous as the change from
the caterpillar to the butterfly. It has often been stated
that the tail and gills of the tadpole *drop off* when it passes
into the adult form: this is incorrect—they are gradually
absorbed. The length of time taken for the whole process of development varies according to circumstances.
Temperature is the most important factor; in warm weather
it may be completed in eight or nine weeks; in cold seasons,
possibly twice as long. Newts, however, are very much
longer than Frogs or Toads in reaching the adult stage
frequently retaining the larval form till the following year.
The differences between the spawn and tadpoles in Frogs,
Toads, and Newts may be tabulated as follows:

	FROGS.	TOADS.	NEWTS.
Spawn deposited..	In shapeless masses	In long strings ..	Singly in leaves, or a few only to-gether in a string.
Tadpoles Develop first ..	Hind Limbs ..	Hind Limbs ..	Fore Limbs.

It is stated that the tadpoles of the Great Warty Newt
are not strictly vegetarians, but will devour insects and the
tadpoles of their smaller cousins. Comparatively few tadpoles
ever reach maturity for they are eaten greedily by the adult
newts, as well as by many water-insects, birds, and fishes,

while Snakes are particularly partial to Frogs at any stage
of growth.   Tadpoles are often spoken of as " Pollywogs,"
by country children.

SUMMARY OF STAGES IN DEVELOPMENT OF FROG TADPOLE.

Hatched without mouth, eyes, gills, or ears.

Develops external gills, mouth, and convoluted intestine ; nostrils, eyes and ears.

Loses sucker under head.

External gills reach limit of size.

Hind limbs begin to appear.

External gills diminish ; Internal gills begin to form.

Hind limbs fully-formed but small ; tail begins to diminish.

External gills vanish ; internal gills fully-formed.

Front limbs form ; tail shortens rapidly.

Intestine becomes shorter and straighter ; lungs enlarge.

Heart becomes three-chambered ; lungs fully developed ;
gills vanish.

COMMON FROG.   Humanity owes a great debt of grati-
Rana temporaria.   tude to the Frog, since to experiments
conducted *in corpore vili,* many valuable
scientific discoveries are due.   No description is necessary
of such a well-known animal, found generally in all
localities suited to its habits.   Like other amphibians it
is nocturnal, and secludes itself during the day in
ditches, or amongst coarse herbage.   Young Frogs often
remain in their native pond after acquiring lungs, especially if the weather be hot and dry, but the first heavy
shower happening towards the end of the day causes
them all to come forth together and start on their
wanderings over the face of the earth.   On such an
occasion, it is no uncommon thing to find a country

Photo by Jones & Son, Ludlow.　　　　　　Specimens at Clungunford.

GROUP OF GREBES, MERGANSERS, AND DIVERS.

1. RED-NECKED GREBE,
2. DITTO (Winter Plumage).
3, 8. SLAVONIAN GREBE (Summer).
9. DITTO (Winter).
11. EARED GREBE,
17. LITTLE GREBE.
18. GREAT CRESTED GREBE,
4. RED-BREASTED MERGANSER (F.)
5, 10. DITTO (M.)
15, 20. GOOSANDER (M. and Y)
21. HOODED MERGANSER
6, 7, 12. SMEW (Y., F. and M.)
13. RED-THROATED DIVER (M.)
14. DITTO (F.)
16. BLACK-THROATED DIVER.
19. GREAT NORTHERN DIVER.

Photo by R. J. Irwin.   Specimens at Hawkstone.

GROUP OF BRITISH GREBES.

1, 2, 3. LITTLE GREBE.
4, 11, 12, 13, 14. GREAT CRESTED GREBE.
5, 6. RED-NECKED GREBE.
10. SCLAVONIAN GREBE.
7, 8. EARED GREBE.

road so thickly covered with little frogs, that it is difficult to avoid treading on them. Country people, noticing the rain and the frogs simultaneously, jump to the conclusion that both came from the same source—hence their accounts of "Showers of Frogs." When on land the Frog progresses by leaps, while in the water it swims by means of the hind-limbs alone. Although the adult Frog is a lung-breather, it can remain a long time under water without coming to the surface for air; when it does come up it may be seen to open its mouth, releasing a large bubble of vitiated air which it had been retaining all this time: it then takes in another mouthful of air, closes its lips, and again dives down. In hot weather it often thus spends hours in the water, although of course it generally lives on land. As the skin is thin and very permeable to fluids, there is no doubt that the blood near its surface is largely aerated through it, as it is through the gills in fishes and tadpoles. The eye, which has a horizontal pupil, is furnished with eyelids, as in lizards; it has also a nictitating membrane. In winter frogs collect together in masses in the mud at the bottom of ponds, etc., "embracing each other so closely as to appear almost one continuous mass." They remain thus in a state of complete torpor, neither feeding nor breathing—unless through the skin as suggested above—but secure from injury, and beyond the reach of frosts: when roused by the return of warmer weather in the spring, they at once set about the important business of spawning, as detailed in the introduction to this chapter. The epidermis, as we have seen, is colourless, but in the under layers of the skin are numerous dark pigment cells to which the colour is due. It has been noticed that the

M

hue of the Frog varies with the locality in which it is found—that in dark places it is dark, while where the surroundings are light, it also is light coloured. This change is believed to be due to the action of light on the pigment cells, which contract when exposed to a strong light, but expand in the dark. The food of the Frog consists of Slugs, Snails, and Insects, which it captures with its tongue in the manner already mentioned. It has many enemies, amongst which Herons, Snakes, Ducks, and Weasels are probably the most destructive. Like the other Amphibia, it rarely comes abroad except by night. When small it has been known to climb up the trunks of trees in search of insects. The Frog is not only one of the most harmless creatures in existence, but exceedingly useful to the gardener in keeping in check slugs and plant-lice (*Aphides*), two of his greatest plagues. For this reason it ought to be encouraged in every possible way, instead of being—only too often—subjected to cruel persecution at the hands of thoughtless boys, and even grown-up persons, who ought to known better. The general colour is brownish, with variable dark markings. Length, 3 inches.

EDIBLE FROG. Eyton in his "Fauna of Shropshire and
R. esculenta. North Wales," written in 1836, states
that this species is found on the Weald Moors. He writes, "During the war some French emigrants who were at Wellington, were highly delighted at finding the true sort in this locality." The writer has never been there, nor has he received any report from that district, but it would be a very interesting task for any resident there to search now and ascertain if the species still survives. "The principal features by which

the Edible Frog may be distinguished from the Common Frog are:—the absence in the former of the conspicuous dark patch, which in the latter extends from the eye to the shoulder: the vocal sacs, or bladders, at the angles of the mouth in the Edible Frog, which are distended while croaking, and which are absent in the Common Frog: and the light line which runs down the back of the former, but which is not seen in the latter. To these may be added the more distinct and beautiful markings in the Edible Frog, its louder note, and generally larger size," *M. C. Cooke.* The hind feet are more completely webbed than in the Common Frog. The ground colour is greenish with black markings. Length, 3¼ inches. The food and habits are similar in both species, but the Edible Frog is found always in, or close to water, and when alarmed always plunges in, and does not again show itself till certain that the danger is past. The Edible Frog is common on the continent, but it is very doubtful whether it is indigenous to our own country, as it was intentionally introduced in 1837 to the Fen Country. What is *most strange*, however, is that Eyton states positively that it occurred *near Wellington, Salop, at that very time.* The French emigrants evidently did not introduce it themselves for he says that they were "highly delighted at *finding* the true sort" there. Eyton was far too good a naturalist to make a mistake as to the species, and we can only conclude that the Edible Frog is either a true Native, or that it was introduced by some person, and at some period unknown. We often hear about the French eating Frogs: they do not eat the whole animal—only the legs, as stated in the following account written by the late Frank Buckland:

" I went (he says) to the large market in the Faubourg St. Germain, and inquired for frogs. I was referred to a stately-looking dame at at a fish-stall, who produced a box nearly full of them, huddling and crawling about, and occasionally croaking, as though aware of the fate to which they were destined. The price fixed was two a penny, and having ordered the dish to be prepared, the *dame de la halle* dived her hand in among them and having secured her victim by the hind-legs, she severed him in twain with a sharp knife; the legs, minus skin, still struggling were placed on a dish; and the head with the fore-legs affixed, retained life and motion, and performed such motions that the operation became painful to look at. These legs were afterwards cooked at the *restaurateur's*, being served up fried in bread-crumbs, as larks are in England; and most excellent eating they were, tasting more like the delicate flesh of the rabbit than anything else I can think of. I afterwards tried a dish of the common English frog, but his flesh is not so white nor so tender as his French brother."

COMMON TOAD.
Bufo vulgaris.

Plentiful all over Shropshire. So deeply rooted is the prejudice against toads that few persons will venture to handle one, even though they know that it is quite harmless. Certainly we cannot plead that it is possessed of any personal beauty, or even that its appearance is at all "pleasant to the eye." Its leaden hue, warty skin, short legs, and heavy thick-set form, inspire only feelings of aversion and disgust. Yet we can truthfully say that the Toad is quite as useful to the gardener as the Frog, in keeping down slugs and insect pests. This fact is recognised by many intelligent gardeners, who encourage Toads to stay amongst the beds, and even fetch them to place in greenhouses and frames. The ugliness of the Toad actually stands it in good stead, for it repels animals and birds that eat its prettier cousin, the Frog. Even more efficient protection is afforded, though, by the

warty protuberances on the Toad's skin; these secrete
an acrid, milky, fluid, which is only slightly irritating
to the human skin, but tastes so disagreeable that any
animal or bird that seizes a Toad, drops it again immediately. This fluid has a sulphury odour and exudes from
the pores when the toad is excited in any way. Many
are the curious ideas current about this humble creature,
such as:—that when angry it spits venom at its enemy;
that it will live for hundreds of years enclosed in a solid
block of stone; that it carries a jewel in its head—an
idea perpetuated by Shakespeare. It is hardly necessary
to say that all these notions are equally unfounded. In
its habits and mode of feeding the Toad resembles the
Frog, but is of a more sluggish disposition, and it crawls
over the ground instead of leaping. It has been seen to
eat earthworms, not taking them with the tongue, but
seizing one by the middle between its jaws and gulping
it down by a series of jerks; the worm meanwhile
struggling to escape, "the toad thrusting it back all the
time, and forcing it down its throat by the aid of its
fore-feet." When intent on taking its prey, the Toad
only shows by its attitude, and by a slight twitching of
the hind-toes, that it is on the alert; it always waits till
the insect moves, thus showing that it is alive; then out
darts the Toad's tongue, and—the insect is gone! So
rapid is the whole action that the eye fails to detect any
movement at all. The Toad will never eat anything
that it supposes to be dead. It casts its skin more than
once a year, generally rolling it up and swallowing it like
a big pill. Except in the spawning season, the Toad is
not nearly so partial to water as the Frog; usually it
lurks beneath stones, etc., during the day, and only

comes forth to feed about sunset. Like the other Amphibia it hibernates in holes in the ground, often in companies, though generally solitary in its habits. We have previously described the assemblies for breeding purposes. The general colour is brownish above with darker markings, whitish below with black spots. Length, 3½ inches; the male smaller. The Tadpole is darker than that of the Frog.

NATTERJACK. This is a rather better-looking Toad than *B. calamita.* the common species, the colours being brighter. It is easily distinguished by the light yellowish stripe down the back. The general colour is olive-brown, darkest on the flanks, and besides the stripe just mentioned, it has small light patches scattered over the body, while the marks and pimples are reddish. The under parts are yellowish with black spots, and there are dark bands on the legs. In Shropshire it has only been recorded at Lutwyche, by Mr. R. B. Benson, and at Broseley, by Mr. G. Potts. There is little doubt that it often passes unnoticed, as few persons take any interest in such humble creatures as Toads. The Author will be glad to hear from anyone who may chance to find the Natterjack elsewhere in Shropshire. In places where it does occur it is often found in numbers. It is much more active than the Common Toad, its hind-limbs are longer, and though it does not leap, it sometimes moves at a pace almost amounting to a run. The feet are only slightly webbed and it does not seek water except in the breeding season. The spawn is in long chains, as in the Common Toad, and it is said to be deposited in June, and to develop rapidly. At this season the Natterjack utters a loud hoarse croak, which is continuous, instead of a succession of short notes. Length, 2¾ inches; hind-leg, 2¼ inches. The male is smaller.

GREAT WARTY NEWT. Provincial name, Askel. This *Molge cristata.* is the largest of our three British Newts, and fairly common in Shropshire. As all the three species are very similar in their nature and habits it will only be necessary to describe these in detail once. As compared with the Frog or Toad, it will be seen that the body of the Newt is elongated; the limbs short and weak; the tail long and strong. In the water the Newt propels itself entirely by means of the tail, which is flattened vertically so as to act as a fin, the motion being produced by waving it rapidly to the right and left. When swimming the legs are laid back and pressed close to the body. On land the Newt uses the legs for progression, but they are only just strong enough to bear its weight; leaping is out of the question, and the pace is little better than a crawl. The tongue is less free and flexible than in the Toad or Frog, and consequently plays a much less important part in capturing prey. The Newt generally seizes its food by darting forward and snapping it up with its jaws. Even in its tadpole stage it is much more carnivorous than the Frog, feeding principally on Water-fleas and tiny insects at first, and, as it grows larger, attacking animals of all kinds, including the tadpoles of other Amphibians, or even of its own kind; if unable to eat them, it nibbles bits out of their tails. The tadpoles of Newts, when they hatch out of the egg, have the external gills in a more advanced state of development than those of Frogs at the same stage, and the gills are

in three branches on each side. As before stated, they
are much longer than Frogs in attaining the adult form,
and retain the gills after all the limbs have grown—see
the Introduction to this chapter for details of the
metamorphoses. Even when the young Newt does at
length leave its native pond and begin its terrestrial
existence it is not really adult. It grows so slowly that
it does not attain its full size till the fourth year, and it
probably does not breed before its third year. The
process of egg-laying in the Newt is usually as follows:
the female places her body with the outlet of her oviduct
upon the flat side of the leaf of some water-plant, such as
Callitriche verna, and with her two hind-legs presses the
leaf against her body; as soon as the egg is laid, she
slightly raises the body to allow room for the folding of
the leaf round the egg; this is effected by bending the
edges of the leaf over, and holding them with the hind
feet pressed upon the egg. In about three minutes the
leaf adheres to the glutinous envelope, and is thus kept
in position. The newt then swims away, but has been
observed to return, as if to make sure that all was right
and gently press the leaf over the egg with her lips.
The above is the usual process, but the eggs are not
always laid singly, while if there are no suitable weeds
available, the Newt will lay its eggs amongst moss or
grass close to the edge of the water. In such cases as
many as eight are sometimes found together. The Great
Warty Newt differs from the other two species in having
the skin covered with warts and pimples, similar to those
found on the Toad. They are of the same nature, their
function being to secrete the acrid fluid with the disagreeable smell noticed in that animal, which, by render-

ing it distasteful, serves as a means of defence from those creatures which would otherwise prey upon it. These glands in the Newt are smaller than in the Toad, and the secretion is therefore much less marked. Many people imagine that Newts spend their lives in the ponds where they are found in the summer; this, however, is by no means the case. It is true they spend a longer time there than the other Amphibians, but they are quite as much lung-breathers as the Frogs and Toads, and the adults quit the water and live on land for quite eight months out of every twelve. They then keep secluded under stones, in cool and sheltered situations, and never venture abroad except by night. They are sometimes found in damp cellars, etc., and are then often mistaken for Lizards. The latter, as we have seen, prefer warm and sunny situations, and would never be found in cellars; besides which their skins are covered with scales—Newts have none. If Newts are kept in an aquarium they will live in it happily enough till the end of the breeding season, but then begin to manifest uneasiness, and will escape from the water at the first opportunity. This instinct is particularly strong in the females, who will die unless they can get to land; the males have been known to survive even when kept the whole year in the water. The male Great Warty Newt undergoes a wonderful transformation in the breeding season, when not only does he become resplendent in gorgeous colours, but develops a high crest resembling a cock's comb, running along the back from the neck to the end of the tail; it is highest over the centre of the body, lower over the hind-legs, rising again on the tail. This crest reaches its greatest perfection about May, or

June, and then diminishes rapidly till, by the time the Newts leave the water, it has disappeared entirely, and there is scarcely any visible difference between the sexes except in size—the female is the larger, and has no crest at any season. Even when living on land the skin of Newts feels soft and moist to the touch, like that of Frogs, and they are equally incapable of enduring prolonged drought. They became torpid in winter, and spend it rolled up, several together in a mass, in holes in the ground. The average length of the adult Great Warty Newt is nearly 6 inches. Colour:—blackish above; orange with black spots beneath. The feet are not webbed, and the toes number four on the fore-legs, five (sometimes six), on the hind-legs. The tadpoles, as well as the adult, cast the skin occasionally, and the latter generally make a meal of it.

COMMON or SMOOTH NEWT. Provincial name, Askel.

M. vulgaris. Considerably smaller than the last and easily distinguished by the smooth skin, devoid of warts or pimples. This species is altogether a prettier animal, and the colouring in the breeding season is really beautiful, especially in the male. At this time he is also ornamented with a crest on the back and tail, festooned along the edge; the female has no crest, and the male loses his before he leaves the water to resume his terrestrial life. The upper parts are generally greenish-brown; the lower, orange, and there are numerous irregular black spots on the body and tail. Some specimens, at their brightest, show beautiful shades of blue and gold on the sides. In the female the colours and spots are more obscure and the under-parts often quite plain. Amongst country folk the belief in the poisonous character of this little animal is so firmly rooted that it is extremely difficult to shake it. Many believe that if anyone catches hold of an "Askel," it will turn round, run up its own tail, and bite him, and that the bite is venomous! There is not an atom of truth in this, for the Newt cannot turn upwards in the way mentioned (though naturally it wriggles about in its efforts to get free), neither is the mouth large enough, nor is it provided with teeth big enough, to bite a human finger. There is a still more curious superstition in some districts; that if one of these creatures is accidently swallowed by a human being, cow, or horse, it will go on living inside the stomach, and rob its host of the benefit of any food that may be taken, so that he, or she, will gradually pine away and die! Strange indeed that anyone in these days can believe such utter nonsense. We doubt if it would harm the swallow*er*: we know it would be bad for the Newt!

PALMATE, or WEBBED NEWT. This species resembles the Common Newt in the smoothnes of its skin, and in general appearance. It is distinguished by its smaller size, and, in the breeding season by the male having the hind-feet completely webbed, and his crest low and simple—not festooned—and higher on the tail than above the body. In both sexes the head is broader and shorter than in the common species, and there is a prominent line running on each side of the back from the head to the hind-legs. The tail, instead of tapering to a point, is abruptly truncate, and usually terminates in a slender thread two or three lines in length. It appears that in districts where it occurs, it often exists in such numbers as to replace

M. palmata.

the Common Newt, which may be present in small numbers, or absent altogether. Mr. J. Steele Elliot reports that the Palmate Newt is extremely numerous in the Wyre Forest, and the author has received it from the neighbourhood of Shifnal. He has not found it near Shrewsbury, but there is little doubt that further search would reveal other Shropshire localities for the species. Average length, 3 inches.

CHAPTER VIII.

FISHES.

IN the same way that most Mammals are restricted in their habitat by their inability to travel except by land, Fishes are restricted to living and travelling in the water. The number of kinds found in Fresh water is very limited; by far the larger number of known forms being Marine. The most interesting point in the natural history of these animals is in connection with their breeding habits. Several of the species recorded in the following pages are generally regarded as marine fish, and it will probably surprise many to find that the Flounder, Shad, and Sturgeon have ever ascended the Severn as far as Shrewsbury; although it is well known that the Salmon comes up to spawn. These fish reside, more or less, in the sea, but ascend rivers to deposit their eggs in fresh water; they usually return to the sea soon afterwards. The young after reaching a certain size, also go down to salt water. This is so with most migratory fish, but in the case of the Eel, the rule is reversed; it spends most of its life in *Fresh* water, and when adult goes down to the *Salt* water, and after depositing its spawn in the deep sea, disappears. The characteristics of fishes are so well-known that it is only necessary to briefly allude to them. They are covered with scales which are appendages of the epidermis. [Sometimes these are deeply embedded as in the Tench and Eel, or take the form of plates as in the Sturgeon and Stickleback.] The limbs are modified into the form of fins, which guide

and balance the body, but the flattened tail is the chief
instrument of propulsion. The fins on each side of the
breast are called the pectoral fins, those on the back the
dorsal fins, on the belly ventral fins, the one behind the vent
anal fin, and the tail the caudal fin. The paired fins—
pectoral and ventral—are those which correspond to the fore
and hind limbs of the higher vertebrates. The body is
covered with slime secreted by glands on the flanks; their
openings form a distinct mark along the sides, known as
the lateral line. Respiration is performed by gills which are
arranged on bony arches, and the heart has only two
chambers. There are no eyelids or external ears. The teeth
are generally numerous and pointed, and occur on the palate
and tongue as well as the jaws; they serve only for prehension. Of the twenty-eight species of Fishes here enumerated,
twenty-two are resident, the rest migratory. The names of
resident species are printed in capitals; migrants in small type.

PERCH. The poet Pope speaks of this handsome fish as
Perca fluviatilis. "The bright-eyed Perch, with fins of
Tyrian dye." It is common in pools
and quiet parts of rivers that are more or less sluggish,
and for its bold biting habits it is beloved by youthful
anglers. In large pools it often attains a considerable
size, and, as it is a voracious fish, it is frequently taken
on live and spinning baits intended for pike. It spawns
about April, and is in best condition for the table in the
winter. The Perch of this decade is said to have de-
generated as much as the trout. In some streams it has
almost died out, and although reckoned one of the hardi-
est of fishes, it has been found, by fish culturists, to be
about the most difficult to bring up by hand. The
Severn is fairly well stocked with Perch; the more so

in the Shrewsbury district, since bush-netting has been greatly restricted in this part of the river. Most of the large sheets of water in Shropshire contain plenty of these excellent "coarse fish."

POPE. Local name, " Jack Ruffe." This fish is something
Acerina cernua. like a cross between the Perch and Gudgeon, having the spiny dorsal fin, and general appearance of the former, with the colour and habits of the latter. It does not grow to a large size. The Ruffe is common in the Severn and is to be found on muddy bottoms, where the stream is fairly deep and slow. The fish is practically useless for the table, being small and rather full of bones; but it is said to be good when pickled.

BULLHEAD, or Miller's Thumb. A somewhat ugly
Cottus gobio. little fish, found in most of our smaller streams, lurking under stones, and, if disturbed, darting up stream to another similar hiding place. The skin is almost devoid of scales, but the back of the head is furnished with sharp processes, which have been known to cause the death of Kingfishers by sticking in the throat and choking them.

STICKLEBACK. Local name, Tittlebat. The common or
Gasterosteus aculeatus. Three-spined Stickleback is also plentiful in our smaller streams, but is very different in habits to the last. The males are very pugnacious and sometimes fight to the death, using their sharp spines as deadly weapons. They are coloured brilliant scarlet and blue during the breeding season, and after a fight the victor glows with redoubled splendour. The body is covered with bony plates instead of scales,

and these vary considerably in number. The Stickleback constructs a nest of water-weeds, etc., in March or April, and this is jealously guarded by the male. Its habits make this fish a most interesting inmate of the aquarium.

TEN-SPINED STICKLEBACK. This species is dis-
G. pungitius.　　　　　tinguished from the last by having ten spines on the back, and by the skin being naked or devoid of scaly plates. It was found by Rev. W. Houghton in ditches on the Weald-Moors.

Flounder. This common flat fish sometimes ascends rivers
Pleuronectes flesus.　　　　for a considerable distance above tidal waters. Years ago, before the time of Severn navigation weirs, and when the barge traffic kept a pretty clear channel, Flounders used to be taken in Shropshire ; and one or two older Shrewsbury residents still relate that in their youthful days they occasionally captured specimens when bottom fishing.

Eel. Common in most pools, canals, and rivers. Large
Anguilla vulgaris.　　　　numbers of "elvers" (as the young are called), may often be seen ascending small brooks. Before a thunderstorm Eels become very active, and what is termed in fishing parlance "run." They will then swim about rapidly near the surface, and have been said even to leave the water and to travel over wet grass to another pool or stream. They can exist a long time out of water. Generally speaking their habits are sluggish. Old writers describe two species of Eel—the Broad-nosed, and Sharp-nosed. It is now known that the former is the female, the latter the male of the same fish. The male generally measures about 18 inches, but the female often exceeds three feet

in length. It has only recently been discovered that Eels spawn in the depths of the sea. The young were formerly described under the name of *Leptocephalus*—a genus containing several species. One species has proved to be the fry of the Conger Eel; another, that of the common River Eel. It is believed that Eels die after spawning, as none but "elvers" are known to *ascend* rivers. Eels live several years in fresh water before they attain maturity and, in their turn, go down to the sea to spawn. An Act of George III. (1777), refers to the taking of "elvers (the brood of eels), which come up the river Severn at certain seasons in immense quantities, and afford great support for the inhabitants of the adjacent parishes and places."

Allis Shad. The Act of George III., alluded to with regard
Clupea alosa. to Eels, is made for "the better Preservation of Fish, and regulating the fisheries in the rivers Severn and Vyrnwy," and contains the following :—"And whereas several sorts of fish not mentioned in the said Acts, particularly Lampreys, Shads, and Twaits, which at certain seasons of the year come up the said river in great quantities, cannot be taken with nets of the length and size of the meshes limited by the said Acts, to the great prejudice of the proprietors of and other persons interested in the said fisheries on the said river, and also of the public . . . . ." The Shad is closely allied to the Herring but is much larger, and is distinguished from it by having one or two dark blotches on the sides.

Twaite Shad. A smaller fish, of similar habits to the last,
C. finta. and to which the same remarks will apply. It is distinguished by the gill

N

rakers being fewer and stouter than in the Allis Shad, and the spots on the sides more numerous.

CARP. These fish grow, it is said, to 15lbs. in weight. *Cyprinus carpio.* They are prolific and long lived, but are of slow growth. Carp are hard to kill, and will live a long-time out of water, bearing transport for a great distance, if carefully packed in wet moss. They are not very often taken by anglers; and, in fact, so far as their edible qualities are concerned, the common English Carp—unlike the cultivated varieties of Germany and America—are scarcely worth fishing for. The more common of the familiar "gold fish" are members of the Carp family, brought to this country from China and Japan. What are called "gold" and "silver" fish are varieties of Carp; and latterly there have been introduced into some English lakes a German orange-coloured fish that runs to about 2lbs. or 3lbs. in weight. These are probably the fish that at one time had a temporary home in the "Dingle" pool, in the Quarry, at Shrewsbury, but although they grew large, they did not breed there. There also exists a golden Tench that is very beautifully coloured. It is understood that the golden Tench and golden Carp are a sort of albino variety of the common Tench and Carp. A pool near Hadnall is stated to contain "gold-fish," which reproduce their species there.

GOLD FISH. *(see Carp.)*

CRUCIAN CARP. Eyton says that this fish used to occur in *C. carassius.* a pit near Cotwall, into which it was introduced from Warwickshire. It is not indigenous but was originally introduced from Germany, and has since become naturalized in the Thames and certain other waters.

BARBEL. As we have only one record, and that unfortunately open
Barbus Vulgaris. to some little question, of the existence of
Barbel in Shropshire, this paragraph appears
in small type. The strong sport-giving Barbel is commonly found
in the Thames and in the eastern waters of this country, where it
sometimes grows to 15 lbs. in weight. It ranks in the angler's rough
and ready category of " Coarse fish," and makes but a poor dish for
the table. Why it is that Barbel are plentiful in the Thames and
Trent, and not, on the contrary, in the mid and lower reaches of the
Severn, we are at a loss to explain, as these two important rivers
have natural characteristics much in common. The fish is remarkable for the strong wattles which depend from its jaws, the larger
pair from the upper and the smaller from the lower. Several
angling writers record the fact that the roe of the Barbel is of a
poisonous nature. The record, to which we have referred, of the
capture of a Barbel in Shropshire is involved in some mystery, but
the instance is believed to be authentic. The specimen is said to
have weighed something approaching 15 lbs., and it was caught a
year or so ago—how we refrain from stating—in the deep still pool
in the Severn at Shrewsbury, known as " Blockley's hole." This is
certainly a place where one might expect to find such a fish ; and it
is more than likely that the one landed was not a solitary stranger.

GUDGEON. Occurs in many parts of the Severn, Vyrnwy,
Gobio fluviatilis. Rea, and most of the tributaries, but
its numbers are kept in check by other
fish preying upon it, especially Pike. It lives and feeds
exclusively at the bottom—a habit common to fish that
are furnished with barbels or beards pendant from the
mouth.

ROACH. A pretty silvery fish with bright red lower fins,
Leuciscus rutilus. found in most streams and pools; often
in abundance. Much esteemed by that
large section of anglers commonly termed, but to an
extent misnamed, bottom-fishers. Roach are somewhat

difficult to capture; they swim in shoals, and ground bait is commonly employed by anglers to gather and confine them to one spot.

RUDD. The Rudd is caught in Bomere and Shomere
L. erythrophthalmus. Pools, near Shrewsbury, and in Fenny-mere, near Baschurch. It is confined to waters of a stagnant character, or lagoon-like expanses, like the Norfolk Broads. It is a handsome fish and not so common as the Roach. The prevailing colour of the Rudd is golden, or silver, with a reddish-orange tint. The body is deeper and flatter than the Roach and the head shorter and more "chubby." The most obvious structural difference between Roach and Rudd, and one which never varies, is to be found in the relative position of the dorsal fin; this in the Roach commences, or originates as nearly as possible over the ventral fins, whilst in the Rudd it begins considerably further back. The name Rudd is obviously derived from its colouring.

CHUB. A thick-headed fish, which has for this reason
L. cephalus. been christened the "loggerhead." It is rather dull-coloured, but has bright red fins. The Chub is perhaps the most common fish in the Severn and tributaries. In hot weather numbers may be seen basking in the sunshine close to the surface, or swimming lazily underneath the willows. The flesh of the Chub is coarse and almost tasteless, and it has a large number of forked bones, so that it is held in little regard by anglers. Izaac Walton says of him, "the French esteem him so mean as to call him 'un vilain,'" but nevertheless he gives a recipe by which the flesh

may be made "very good meat." "Red Spinner," a
well-loved authority of the present day on angling, charit-
ably writes, in one of his delightful works, "I have never
yet sought in vain for some poor person who was glad to
receive, and ready to eat, even the logger-headed cheven."
It is recorded that a Chub, "said to weigh 9lbs," was
once taken in the Severn.

DACE. This pretty fish is a general favourite with anglers,
L. Vulgaris.   especially fly fishers, as it rises freely
                 to the artifical fly, and is not bad eat-
ing. It is pleasant to see the "silvery dace" sporting in
large numbers in the Severn, Rea, Tern, Teme, and other
rivers in the County. A correspondent to *The Field* wrote
on December 31st, 1898: "I have been an angler for over
60 years, and have never caught or seen a Dace of 1lb.
The two largest I ever captured were taken with artificial
minnow. The one was 15ozs. and other 14ozs. The
larger one was captured twenty years ago, in the Wylie,
in Wiltshire, and the other in the Rea, in Shropshire."

MINNOW. A pretty little fish, found abundantly in the
L. phoxinus.   Severn and in most tributary streams.
                 It is not generally known that, during
spawning time, male minnows assume very brilliant
hues. "Red Spinner," of *The Field*, tells us he has seen
them "all the colours of the rainbow." This transformation occurs about March. Minnows are in great demand
among anglers, who use them as baits, on spinning tackle,
for the capture of Pike and Trout.

TENCH. A dull-coloured, mud-loving fish, with small
Tinca vulgaris.   deeply-embedded scales, and a thick
                 coating of slime, reputed to have heal-

ing powers with other fish, though this is questionable. It is found in many ponds, especially such as have muddy bottoms.

BREAM. A bright silvery fish with a very deep laterally-
Abramis brama. compressed body, like "a tin plate set on edge." It is found in Berrington, Fennymere, Hawkstone, and other pools, as well as in the Severn, and usually swims about in shoals. It has been also likened to a pair of bellows, and humorously stated to possess much the same flavour! After this it need hardly be added that it is very poor eating.

BLEAK. A small fish, greenish above and silvery white below;
Alburnus lucidus. not uncommon in the Severn, near Shrewsbury, and in certain brooks running into it. Bleak swim in closely packed shoals. The colouring matter of its scales is used in the manu-facture of artificial pearls.

LOACH. Local name, Stone Loach. A tiny little fish with
Nemachilus barbatulus. six barbels on the upper jaw. It gets its name from the habit of lurking under stones, like the Bull-head, and is found in similar situations. It dies in a few minutes if taken out of water.

Salmon. It is a matter of history that this prince amongst
Salmo salar. fishes was formerly caught in numbers in the Severn, close to Shrewsbury, but the pollution of the river, and the excessive netting of the lower reaches, combined with other changes in the channel, has completely altered the state of affairs, and now comparatively few fish pass up. The Salmon is migratory in its habits; it ascends rivers to spawn and,

when there is a sufficient flow of water, will leap over
rocks, weirs, and other obstacles that, alas too commonly,
bar its progress. In different stages, Salmon are known
by various names: samlet or parr, smolt, grilse,
"red-fish" (male salmon at spawning time), "kipper,"
shedder, gilling, botcher, etc. The fry of Salmon (locally
known as samlets), remain one, two, and, in some cases
three years, in the rivers before going down to the sea.
Samlets and the fry of Trout are very similar in appearance, but the former are usually more silvery, and their
spots more uniform in arrangement. The most distinctive feature, however, is in the tiny adipose fin—the
fatty growth behind the dorsal fin. When this fin is
tipped with red or orange it almost invariably proclaims
the fish to be *Salmo fario*, (Common Brown Trout). If these
colours are absent it is wise for anglers to conclude their
capture to be one or other of the migratory salmonidæ,
and to return it to the water, as the capture of salmon
fry is, of course, illegal. There is reason to fear, however,
that large quantities of samlets are, in open disregard of
the law, taken from the Severn and tributaries, by un-
scrupulous anglers, and by poachers, who sweep the fords
with small-meshed nets. It is a well known, and much-to-
be-regretted fact, that these delicate fish find a ready
private "market;" were it not so, poaching of the kind
indicated would not be so prevalent. The capture of
samlets is a most reprehensible practice; the more so
because there are so many causes at work depleting the
Severn and like rivers of Salmon. Attention should
certainly be directed by Parliament to the serious condition of the salmon fisheries of the Severn and other
important inland fisheries. "Severn Salmon" is now an

expensive luxury; but it bids fair to reach a price
prohibitive to most consumers. A prominent angling
authority wittily wrote, on a recent occasion, that the only
taste of Salmon which the general public have is through
the medium of a foreign tin. The difficulty of absolutely
determining to what species the fry of migratory salmonidæ actually belong is sometimes considerable, and the
subject frequently gives rise to controversy. An old
work on angling by C. Bowlker, of Ludlow, contains a
paragraph concerning fish to which the peculiar name
of "Gravel Last-Spring" is given. This runs as follows:
"The Gravel Last-Spring is supposed by some to be
the fry of Salmon, but which is a distinct species; the
rivers Severn and Wye abound with this fish. It spawns
in the month of August, and affords the angler excellent
diversion with a long line. The Red Ant is a very killing
fly, and all the flies may be used with success during
their proper seasons." Authorities are now agreed that
these "Gravel Last-Spring" are purely samlets, but there
are Severn anglers who yet maintain that such is not the
case. Some, like Bowlker, aver that the little fish belong
to a distinct species, and they support their contention by
stating that they have caught female specimens, only a
few inches in length, having roe pretty well developed.
In this connection the following may be quoted from the
Rev. W. Houghton's work on Fishes:—"The female
Salmon is mature when about fifteen inches long; the
male, however, may be mature when in the smolt stage,
and about six or seven inches long." The Severn Con-
servancy Board forbids the capture of so called "Gravel
Last-Spring," holding of course that they are indubitably
the young of Salmon. The principal migration of mature

Salmon to the upper districts of the Severn and like rivers, for the purpose of spawning, occurs during the first floods of winter, and on into December ; but a few gravid fish continue to run through the early months of the next year. As therefore mature Salmon arrive at the higher reaches of rivers at different periods, some are later than others in spawning, and the fry are consequently later in hatching out. The Rev. W. Houghton says, "Parr hatched in February, 1878, may be ready to take their journey seawards on May or June, 1879. It has also been shown that a large number remain in the river till they are a little over two years old; a parr hatched in February, 1868, may remain in the fresh water till May or June, 1870, and I suspect that this is usually the case." The principal migration of parr (or samlets), to the sea occurs about April. The Parr, putting on a silvery dress, as a sign that its time for seeking salt water has arrived, is called a Smolt. When captured at this time the old scales come off in masses very easily. The fish will not undertake their journey to the sea until the state of the water is favourable, and they are, moreover, in a fit condition to depart. It is held by many ichthyologists that if they are not ready for migration in the spring they remain in the river during summer, and it is, in all probability, to these stragglers that the name of Gravel Last-Spring, above referred to, is given. The appellation may be accounted for by the fact that the fish are the last to spring from, or leave, the gravelly spawning beds. There is a remote possibility that the "Gravel Last-Spring" are the fry of Sea Trout ; but this is not very likely to be the case.

Sea Trout, or Salmon Trout. The fact that Sea Trout
Salmo trutta.   are included in our list of Shropshire
                                   fishes may cause some surprise, even
to experienced Severn anglers; nevertheless these fish
were at one time frequently taken both in the Severn
and in the Vyrnwy. This statement has been received
with some incredulity, but careful inquiries have resulted
in reliable testimony as to its verity, being obtained, from
several sources quite independent of each other. The
Rev. J. B. Meredith reports that he has several times
taken Sea Trout in the Vyrnwy, near its confluence
with the Severn. On one occasion the rev. gentleman,
and the late Mr. Harry J. Potts, were angling in the
Vyrnwy, and the former, in his friend's presence, landed
a lively fish of about 1lb. in weight. Mr. Potts immediately exclaimed, " Why, it's a Sea Trout," and a close
examination confirmed his first impression. This gentleman had had considerable experience in angling for
Sea Trout, in various parts of the country, and Mr.
Meredith—who had long believed that these fish migrated
to the upper districts of the Severn—considers the proofs
he has given absolutely incontrovertible. Several
Shrewsbury gentlemen have since stated that they have
either caught Sea Trout, or witnessed their capture, near
the town ; but no recent instance is recorded. It would
therefore seem that the migration of these fish to the
upper Severn and Vyrnwy is now rare. Like all navigable rivers in this country, the Severn has undergone
material changes, and when we endeavour to picture the
stream as it once existed, it is no excess of fancy to
imagine it to have been well stocked with migratory
fishes—even above Shropshire—at certain periods of the

year.   In days long gone by, Salmon and their kind, had
an almost entirely uninterrupted passage from the sea to
the spawning grounds, in the rivers Vyrnwy and Severn,
in the counties of Salop and Montgomery; comparatively
few nets were set, or used in the estuary to catch the
fish as they "ran;" chemical and sewage pollution did
not then cause them to turn back as it is now supposed
to do; the ancient barges kept free the channel
for the travellers; and lastly, there were few weirs or
other obstacles to check or bar their progress.   It is
partly owing to these changes, but more particularly to
the excessive netting which takes place in the estuary,
that Salmon are now so scarce in our locality; and a
similar state of things has been brought about in other
rivers.   It is therefore not unreasonable to conclude
that, as Salmon have practically disappeared from the
higher reaches of the Severn, Sea Trout, through like
causes, are not now to be found in the rivers in this
County.   The following description of Sea Trout, by the
Rev. W. Houghton, should enable anyone who captures
one in local waters or elsewhere, to identify the
fish :—"Colour of the body above lateral line, dark-
bluish, lighter on sides, belly silver white, the black X
shaped spots on this fish are generally very distinct, for
the most part they are above the lateral line, but occasionally there are a few below; the gill cover is usually
marked with a few round black spots; adipose fin dark,
free from any red tinge ; the scales are round and small,
and easily detached; tail nearly square."   The same
author describes Sea Trout as being "great and pertina-
cious travellers," and quoting from Günther, he says
they reach a length of about three feet.

TROUT. No fish is a greater favourite with anglers than
Salmo fario. Trout, of which there are several
varieties. The Common Brown Trout
(*Salmo fario*), is the only one natural to the Severn and other
local streams. The Shropshire Angling Association
has however turned into the Severn at Shrewsbury, and
also into some of the tributaries, a quantity of Loch Leven
Trout (*Salmo levenensis*), and these seem to have thriven, as
good specimens have from time to time been taken. It
is more than probable that these Loch Leven fish will
not retain their distinctive features, but will assimilate to
their new locality. The Common Brown Trout was once
pretty plentiful in the Severn, but, like Salmon, it is
becoming much scarcer, in consequence of over fishing
and the depredations of Pike.

RAINBOW TROUT. This beautiful species is of a bright golden
S. irideus. colour and usually has a pink band on the side.
It is an American fish, but was introduced
recently into a private stream at Bourton, where, it is stated, to
be increasing in size and thriving well. Two specimens have been
taken from the Severn near Shrewsbury, with a rod and line, so that
the fish have apparently worked their way down the brook into the
main stream.

GRAYLING (or Umber). One of the most delicate and
Thymallus vulgaris. symmetrical of fishes, excellent for the
table, and highly esteemed by anglers.
It affords sport to the fly-fisher when Trout and Salmon
are out of season, namely, in the autumn, winter, and
early spring. Grayling are fairly numerous in the Severn
notably at The Isle, Atcham, Cronkhill, and other places
where the bed of the river is gravelly, and there is a
gliding current. The Teme and several rivers in the

Ludlow district are among the best Grayling-producing rivers in the kingdom, and anglers travel far, and pay high prices, for the fishing rights in them—especially at Leintwardine and the immediate neighbourhood. A characteristic feature of the Grayling is its great dorsal fin, which, if held to the light when the fish is freshly caught, exhibits the most lovely colours. The newly captured Grayling emits a very perceptible odour, not unlike the scent of thyme. Izaac Walton says: "some think that he feeds on water thyme, and smells of it, at his first taking out of the water."

PIKE, or Jack. The "pet aversion" of the salmon, trout,
Esox lucius. and grayling angler, on account of the havoc wrought by him among these fish and their fry ; no river fish being safe from his attacks. Many are the stories told of the voracity of the Pike, for, not content with preying on other fish, he has been known often to devour ducklings, water rats, and any other "small game" he happens to come across. It is known to live to a great age and attain to a weight of 30lbs. or more. Pike have greatly increased in the Severn and Vyrnwy of late years—unfortunately for the well-being of Trout fisheries in these rivers and in the tributaries.

Sturgeon. Though a marine fish, the Sturgeon ascends
Acipenser sturio. rivers for a considerable distance to spawn, and on Sept. 12th, 1802 (Eyton gives the year as 1799) a fine specimen was caught in the Severn below Shrewsbury Castle, and its skin placed in the Museum. It was 8½ feet long and weighed 192 lbs. The Sturgeon is interesting as belonging to an order of

fishes (Ganoids) which were the predominant group in the seas of a long past geological period. They differ much from the fishes mentioned in the preceding pages in many respects, and there are few representatives of the order existing at the present day. The most obvious differences in the Sturgeon are the form of the scales and tail. Instead of having scales overlapping like roofing-slates, the Sturgeon has conical bony plates arranged longitudinally on the body, with bare skin between them, and the tail has the termination of the backbone prolonged into its upper lobe. "Caviare" is made from the roe of the Sturgeon of Northern and of Eastern Europe, and isinglass from its swim-bladder.

CHAPTER IX.

THE LOWEST VERTEBRATES.

THE Lampreys belong to the very lowest group of vertebrates. They are not regarded as " Fishes " by modern zoologists but present so many peculiarities that they form a sub-class by themselves. In the first place they have no gill-arches such as are found in fishes, but, instead of these, are provided with 6 or 7 gill *pouches* on each side. They have no ribs or limbs and the skull is immovably fixed to the backbone; the skeleton is cartilaginous, and the skin bare of scales. The mouth is circular and adapted for sucking, but when closed is only a slit. It is provided with horny teeth which rasp off the flesh of any fish to which the lamprey attaches itself. Unlike fishes, too, lampreys undergo a metamorphosis and take three or four years to reach maturity. The young are so unlike the adult that they were regarded as distinct animals under the name of *Ammocœtes* until their true nature was discovered by watching their metamorphoses. In spite of all these differences Lampreys bear a close resemblance to Eels and have a fin running along the back as in those fishes.

Sea Lamprey. This is the largest of the three species, *Petromyzum marinus.* reaching to a length of 3 feet, and is further distinguished by its colour— greenish brown or grey mottled with black. In the adult state it resides in salt water but always ascends rivers to

spawn; the Severn in particular has long been celebrated for its Lampreys. This species does not often
now occur as high up as Shrewsbury, though it was not
uncommon in former years. One was caught below the
Welsh Bridge a year or so ago, and a specimen in Ludlow
Museum was taken in the Teme at Tenbury, July 5th,
1878. This is undoubtedly the Lamprey referred to in an
Act of George III. (*vide* Shad), and it is the fellow that
is reputed to have caused the death of Henry I. Dr. Day
in his "British Fishes" figures one of the Severn lampreys
as 36 inches long. The fact that the Sea Lamprey is
getting rare will go to excuse the lengthy reference to it
herein, as in another decade or so it may become quite
extinct, at any rate in the waters of the Severn. Moreover but little appears to be known of it in this district,
although there is every reason to believe that it was
plentiful in Shropshire and Montgomeryshire before the
construction of navigation and other weirs prevented its
passage from the sea. The opinion may be hazarded that
it is now becoming rare because the breeding Lampreys
cannot reach the quieter and more favourable spawning
places out of the way of launches and other disturbing
conditions; owing to which the spawn is rendered
unfertile. The *Bradford Observer* in an article published
towards the close of 1898 gives some interesting facts
concerning Lampreys, portions of which are here
quoted. Although it is essentially a marine species, the
great Lamprey occasionally penetrates from the usual
estuarine breeding grounds far into the fresh-water rivers,
beyond the tidal influences. In a tributary of the Severn
they are sometimes seen during the month of May,
engaged in operations of a somewhat peculiar nature.

Having attached itself to a stone of a suitable size, the creature swims to the place selected, supporting the weight with infinite skill until the moment when it suddenly drops it into the muddy bed of the stream. Then the stone is worked backwards and forwards for some hours, the discolouration of the water serving to reveal the presence of the Lamprey from the bank. After the task has been accomplished by these incessant efforts, a suitable excavation remains for the deposition of the eggs in the required position. The mouth of the Lamprey during the alternations of expansion and contraction is a truly formidable looking cavity. A freshly caught and still living specimen has the power of emitting a strange hissing sound which is doubtless due to the contraction of the sucker-mouth. The Sea Lamprey sometimes falls a prey to the attacks of numerous parasitic leeches adherent to its body by means of suckers—a curious case of retributive justice, since the Lamprey is itself parasitic on Fishes.

Lampern, or River Lamprey. Local name, Seven eyes, *P. fluviatilis.* in allusion to the seven gill-slits. Distinguished from the last by its smaller size—never over fifteen inches—greenish-blue colour, and the absence of mottling. It is common in our Shropshire streams, where bunches of eighteen or twenty may often be seen lying in holes in the gravelly bottoms, which it prefers for spawning. This used to be considered a purely freshwater form, but has been taken in the sea, and is now known to migrate in the same way as the Sea Lamprey. The young are believed to pass through their metamorphoses and reside in fresh water for three or four years.

Mud Lamprey, or Sand Pride.   The smallest of the three
P. branchialis.   species, rarely exceeding ten inches in
                                    length.   It closely resembles the last
in habits and appearance, and is common in our brooks.
It is said to die directly after spawning.   Sea fishermen
use it commonly for bait, for, as in the other species, the
adult is found in the sea.

THE END.

INDEX.